LAIR OF THE LION

LAIR OF THE LION

A History of Beaver Stadium

Lee Stout

Harry H. West

The Pennsylvania State University Press, University Park, Pennsylvania

A KEYSTONE BOOK®

Keystone Books are intended to serve the citizens of
Pennsylvania. They are accessible, well-researched explora-
tions into the history, culture, society, and environment of the
Keystone State as part of the Middle Atlantic region.

Library of Congress Cataloging-in-Publication Data
Names: Stout, Leon J., author. | West, Harry H., 1936– ,
    author.
Title: Lair of the lion : a history of Beaver Stadium / Lee
    Stout, Harry H. West.
Description: University Park, Pennsylvania : The
    Pennsylvania State University Press, [2017] | "Keystone
    books." | Includes bibliographical references and index.
Summary: "Describes the evolution of Penn State's Beaver
    Stadium (originally Beaver Field) and its iconic status
    for the Penn State community. Traces the history of the
    stadium within the context of the university's history and
    explores how fans have experienced football games from
    1887 to the present"—Provided by publisher.
Identifiers: LCCN 2016046594 | ISBN 9780271077765 (cloth
    : alk. paper)
Subjects: LCSH: Beaver Stadium (State
    College, Pa.)—History. | Pennsylvania State
    University—Football—History.
Classification: LCC GV417.B43 S76 2017 | DDC
    796.406/80974853—dc23
LC record available at https://lccn.loc.gov/2016046594

The Pennsylvania State University Press is a member of the
Association of American University Presses.

It is the policy of The Pennsylvania State University Press to
use acid-free paper. Publications on uncoated stock satisfy
the minimum requirements of American National Standard
for Information Sciences—Permanence of Paper for Printed
Library Material, ANSI Z39.48–1992.

TO our wives, Dee Stout and Laurie West, who joined us in sharing the ups and downs of this project with good cheer and strong encouragement

& TO the men who happily shared their love of Penn State football on their visits and contributions to the Penn State Archives: Tiny McMahon, Charley Way, Jack Sherry, Rip Engle, Ridge Riley, Wally Triplett, Roosevelt Grier, and Carl Stravinski

& TO the civil engineering students who proved that playing football at Penn State did not interfere with achieving academic excellence and professional success: Neal Smith, John and George Kulka, Craig Lyle, Brian Siverling, Ralph Giacomarro, and Scott Shirley

CONTENTS

ACKNOWLEDGMENTS

This book began with a discussion in a State College gym in 2011. Professor Harry West was teaching a course on the history of structures, which included a lecture on Beaver Stadium, followed by a behind-the-scenes tour of the facility. He had expanded the class session into a program that was well received by student groups, local organizations and alumni clubs. Between exercises, he suggested to Lee Stout that it might make an interesting book. West had read Stout's "Penn State Diary" columns in *Town & Gown* magazine and his book *Ice Cream U* and thought that he might want to collaborate.

In the spring of 2011, Stout attended the Beaver Stadium lecture and took the tour; he quickly realized that this project had real potential. However, with so many Penn State football history books on the market, there was no point in just retelling the stories of great teams, seasons, games, or individuals. Stout concluded that, as a framework for describing the engineering and construction history of the stadium, the book should explore the historical context for football at Penn State and its meaning for the institution, the students, and the alumni. It should also examine the game-day experience of the fans at Beaver Stadium and how that changed over the years. West agreed that this approach would provide an interesting blend.

They were developing a draft proposal near the end of the 2011 football season when the Sandusky scandal, Joe Paterno's death, the Freeh Report, and the NCAA sanctions derailed the project. They wondered if there would still be an audience for a book that traced the history of the venue and the thrill of attending a Penn State football game and, as these events unfolded, they set the proposal aside.

Fortunately, the Penn State football team reacted to the scandal and the sanctions with spirit and integrity. Fans of the Nittany Lions responded with pride in their team and refused to abandon Penn State football. After Bill O'Brien's gritty first season as head coach, the Penn State football program appeared to be surviving despite all the challenges it faced, and the authors resumed working on the book.

A proposal and sample chapters were ready by the fall of 2013 and the authors brought them to the Penn State University Press early the next year. The Press liked the idea and work began in earnest. By the fall of 2015 a complete manuscript was ready for expert review. After revisions, they completed a final draft of the manuscript, with illustrations and figures, in early 2016. Since then the editorial, design, and production process has wound its way to completion and the book you now have in your hands is the result.

To a large extent, this collaboration is the product of many years of research and teaching structural engineering on the part of Harry West, and archival work and writing about Penn State history by Lee Stout. But such a project would have been impossible without the assistance and contributions of many others.

Over Professor West's years of teaching about Beaver Stadium, he benefitted from the help of W. Herbert Schmidt, former associate director of athletics; Chris Yarger and Kenneth Johnson, both from the Office of Physical Plant; Robert DuPuy Davis, consulting structural engineer, formerly of John C. Haas Associates; Robert Barnoff, structural engineer and former head of the Department of Civil and Environmental Engineering; and Kenneth Getz, former job superintendent for Alexander Construction Company.

In preparing the book, we received assistance from and spoke with numerous individuals about stadium operations and history. These included Bobby White, director of marketing operations for suites, club seats and special events for Beaver Stadium, and Brad "Spider" Caldwell, his assistant director; Tim Curley, former director of athletics; Bud Meredith, former ticket manager; Eric and Robert Byers, former assistant managers of concessions; Mark Kresovich, stadium turf crew leader; Professor Emeritus of Turfgrass Management Alfred Turgeon; Steve Jones, Penn State sports announcer and instructor in the College of Communications; and Penn State President Emeritus Graham B. Spanier.

Several experts on the history of Penn State and athletics read the manuscript at various stages and provided important feedback. They include W. Herbert Schmidt, Michael Bezilla, former University historian; Roger Williams, former executive director of the Penn State Alumni Association; Ron Smith, professor emeritus of kinesiology and sports historian; John Black, editor of the *Penn State Alumni Association Football Letter*; and Lou Prato, sports historian and former director of the Penn State All-Sports History Museum. Their insights and suggestions helped enormously; any faults and errors in the book remain our own. In addition, our wives, Dee Stout and Laurie West, read every draft and version of the manuscript and provided suggestions that improved the final product at every turn.

At various stages of manuscript preparation, we had the assistance of Paul A. Bowers, assistant professor of architectural engineering, and draftsperson Amy Pennebacker, who were responsible for the book's various technical figures; and Roxanne Toto, instructional designer in the College of the Liberal Arts, who helped on illustration enhancement. Jackie Esposito, Paul Dzyak, and Paul Karwacki of the Penn State University Archives and its Sports Archives were unfailingly helpful in providing information and illustrations. David F. McCartney, University Archivist at the University of Iowa, Kathleen O'Toole, lecturer in the College of Communications; Mary Sorensen, executive director of the Centre County Historical Society; and photographers Greg Grieco and Steven P. Benner all provided resources and materials of value in creating this book, as did Mark Zehr of ProCopy in State College and Mark Raffetto of the Penn State Engineering Copy Center, who helped with our image challenges.

Naturally, we thank the staff of the Penn State University Press, including director Patrick Alexander, Kathryn B. Yahner, Jennifer Norton, Laura Reed-Morrisson, Hannah Herbert, Patty Mitchell, and Regina Starace, along with our independent copyeditor Therese Boyd, all of whom provided vital guidance and assistance in the creation of this book.

Finally, we must acknowledge the students, colleagues, alumni, friends, and Penn State fans whom we have encountered over the years for adding to our knowledge, testing our abilities to tell these stories in an understandable fashion, and for inspiring us to persevere in this project despite the distractions and disappointments over the years since 2011. The recovering pride of the Penn State community helped make this project a reality.

INTRODUCTION

The car is packed, you have all the food and other tailgating supplies, everyone's in, the parking pass is hanging from the mirror, and you've even remembered the tickets. It's time to head to Beaver Stadium for another Penn State football game. No matter how long the drive, there's a tingle of anticipation for another full day of football and fun.

More than 100,000 fans—students, alumni, and Penn State faithful from all over Pennsylvania and surrounding states—gather for a festival that is focused on a football game. It is said that on a home game Saturday, this corner of the Penn State campus becomes the fourth-largest city in the Commonwealth. Ascending to the upper reaches of the stadium provides a view that makes that assertion completely believable.

But the events of the day include more than just the game. We look forward to tailgating, cheering the players and the Blue Band as they arrive, and then immersing ourselves in the sights, sounds, and smells of the stadium. We watch the game, cheering and singing along with the band and the popular anthems, like "Sweet Caroline" played over the stadium sound system. We brave the crowds for halftime refreshments and relief, and the contest resumes. Eventually the game ends, the team and the student section sing the alma mater, the players ring the victory bell on their way to the locker room, and the band returns to the field for its postgame show. Then it's out to the parking lots and more tailgating. Finally, we pack up and head home.

Being a part of this daylong celebration at Beaver Stadium is an unforgettable element of the Penn State student experience. And as students move on into adult life, days like this will be "re-experienced," as alumni return for games throughout their adult lives. The game ties us to the University; its traditions and customs become cherished memories. Central to those memories are the place, the stadium and the surrounding grounds.

In the chapters that follow, we will trace the evolution of the two Beaver Fields and Beaver Stadium, and describe the engineering challenges encountered along the way. In addition, we will examine what the football game-day was like for fans at different times in the past and review both the tangible markers of history and intangible customs and heritage that permeate the stadium. We will also consider the changing traditions and rituals, and the significance of football as a sport at Penn State. From tailgates to turf management, from refreshments to music, cheers, and the deafening roar of "We are . . . Penn State!" we'll attempt to provide the historical background for it all.

If Beaver Stadium is truly the "Lair of the Lion," then we hope that you, the reader, will come away with a true appreciation for how this iconic venue "got to be that way."

OLD BEAVER FIELD

The inauguration of Penn State's new outdoor sports field and grandstand was scheduled for Saturday, November 4, 1893, with a football game against the Western University of Pennsylvania (today's University of Pittsburgh). But as the *Free Lance*, the predecessor to the *Collegian*, reported, "When the boys from Allegheny arrived, they found the vilest kind of weather and agreed to postpone the game till Monday."

The visiting players were graciously hosted for the weekend by the college's fraternities, and the cordiality persisted despite a 32–0 Penn State victory. The *Free Lance* noted "the marked absence of slugging. Both the teams behaved like gentlemen and the most friendly relations existed between the players."

That Monday afternoon game was played in "perfect football weather," and it was the Nittany Lions' only home game of the five played in the 1893 season. General Beaver and his wife were in attendance, as they usually were for college events. As governor, he had secured a legislative appropriation to fund the grading of the field and the construction of the grandstand. The students and trustees agreed that the field should be named for Beaver, recognizing his many years of service and support for the college and its students.

Today's Beaver Stadium, with its 106,572-seat capacity, is a far cry from Old Beaver Field (as we call it today), where 500 fans could sit to watch track meets, football, and baseball games in the center of the Penn State campus. It's hardly the only thing that has changed in Penn State sports since the nineteenth century; both the football game and how fans experience it have also evolved in significant ways.

FIGURE 1   President George W. Atherton

scientist, ended abruptly with his unexpected death in 1864. Five successors tried in a variety of ways to make the college an effective and financially sound institution, but with no great success. By 1880 Penn State was on the verge of being stripped of its land-grant endowment, which would likely have forced it to close its doors. Then, in 1882 a new president, George W. Atherton, arrived; he would be the spark for a revitalized Penn State and would start to build the "great industrial university" that Pugh had foreseen in the 1860s.

Atherton was an 1863 graduate of Yale, later the nation's premier athletic school. While he had been a classical scholar, by the time he arrived at Penn State he was the nation's foremost proponent of a national system of land-grant colleges. Atherton understood the ideals of land-grant education perhaps better than any other educator in America. After teaching at the University of Illinois, he was for thirteen years Voorhees Professor of Political Economy and Constitutional Law at Rutgers University (where he may well have observed the first college football game in American history). He was admitted to the New Jersey bar and served on several federal and state government commissions.

Atherton had more political acumen than any previous Penn State president, and he was a young and vigorous man in his forties. He was coming to a school where scientific and technical education could potentially be a partner to support the growing industrial power of the Commonwealth of Pennsylvania. Within his first decade of service, George Atherton had changed Penn State's course and fortunes. New curricula—especially a much expanded engineering program—more than tripled enrollments and the legislature responded for the first time with regular appropriations, especially for new buildings.

Nevertheless, it all begins with students watching their classmates play a game, in a space on campus created for that competition and its spectators.

The Penn State Context for Sports

Penn State, founded in 1855, went through a difficult birth and childhood. Pennsylvania's land-grant college's first quarter-century included years of both promise and futility. The visionary leadership of the first president, Evan Pugh, a highly respected

Atherton's partner in promoting Penn State's interests in Harrisburg was Bellefonte lawyer and college trustee James A. Beaver. He was elected governor in 1887 and, although he was not an alumnus, he had sincere affection for the school and its students. Beaver had led a regiment in the Civil War, lost a leg in combat, and earned the rank of brigadier general for his services. After his gubernatorial term, he was named one of the first justices of the new state superior court. Penn State had been a family project; his father-in-law, Hugh Nelson McAllister, was one of the college's original trustees. Beaver succeeded McAllister in 1873, serving on the Board of Trustees until his death in 1914, including twenty-four years as board president. During his term as governor (January 18, 1887 to January 20, 1891), Beaver was particularly successful at securing legislative appropriations for new buildings, including the athletic field that was named for him. Today's Beaver Stadium continues that recognition and honor.

At the same time that Penn State was struggling to survive in the 1850s and '60s, student life at our colleges and universities was in the early stages of a significant change. The paternalism of the old-time college president and his faculty was beginning to fracture. Outside the classroom, students were taking charge of their lives. Starting in the 1880s Penn State students began to find outlets in social fraternities, through publications and the arts, and in sports.

These activities made up the "extra-curriculum," and student initiative and energy drove them as administrative control of student life gradually eased. Sports were the most popular component of those activities. Initially, baseball was the favorite sport, but football soon displaced it in the affections of the student body. In 1893 the *Free Lance* asserted that, with the opening of Old Beaver Field, "the game is

FIGURE 2 General James A. Beaver

here to stay. It has fastened its hold on the mind of the younger generation and it is bound to thrive."

However, it was not just students who were enthusiastic about athletics and football. Soon the growing numbers and involvement of alumni and locals began to strengthen support for sports. A college alumni association was established in 1870, even though there had only been a total of 60 degrees awarded by the 1869–70 academic year. Nevertheless, at the urging of General Beaver, three seats elected by the alumni were added to the Board of Trustees in 1876 and expanded to nine alumni trustees in 1905. Significantly, some of the most important alumni trustees of these years had been athletes during their undergraduate years. J. Franklin Shields, later board president (1929–46), was football team manager in 1891 and the moving force behind the short-lived Pennsylvania Intercollegiate Football League.

Even some of the faculty and administrators shared pride in athletic achievement. President Atherton felt football was good for college spirit and his son Charles became a star player for Penn State from 1890 to 1894. English professor and Dartmouth alum A. Howry Espenshade lamented, however, that he had "never known a college where athletics were more highly favored and encouraged, particularly by the faculty, many of whom, like the students themselves, ranked successful athletes far above capable students, of whom we had too few."

Like Espenshade, many American professors found the growing fervor for sports inappropriate for institutions of higher learning. There were outspoken critics among American college faculty members, who were especially troubled by the brutality of the game of that era, the prevalence of betting on games, and the falling away of the amateur ideals with the appearance of increasing professionalism and commercialism. Individual college presidents and faculties varied considerably in the degree of control they chose to exert over athletics in their own schools. Eventually, these concerns resulted in the creation of the Intercollegiate Conference of Faculty Representatives (the founding body of the Big Ten) in 1895 and the National Collegiate Athletic Association (the NCAA) in 1905, both of which tried to bring some measure of uniform inter-institutional control to football and other sports.

Students Create an Athletic Program

The *Free Lance* is probably the most colorful record of the rise of intercollegiate athletics and the eventual development of facilities for them at Penn State. Starting publication in the spring of 1887, the paper was an advocate for sports on campus and a reporter of college sports elsewhere. The accomplishments of students at Yale, Harvard, and various Pennsylvania colleges, and the gifts of those schools' alumni for gymnasiums and sports fields, presented an implicit contrast to Penn State, where fundraising and football (except for one game in 1881 against Bucknell) were nonexistent.

The ineffectiveness of the student athletic association, its "lack of energy and pluck," was also lamented, and regular cries of "Wake up!" were voiced in the pages of the *Free Lance.* But without facilities, little could be done. In October 1887 the paper reported, "the ball-ground has been rendered useless," and the editors were pessimistic that a football team would be formed. "Why don't the College Authorities show a little more interest in the sports of the P.S.C.?" they grumbled.

Like their peers on other campuses, nineteenth-century Penn State students were active young men and women who increasingly channeled their loyalties toward their school, in addition to the traditional affection for their college class and close friends. School colors, mascot, songs, and cheers became common accompaniments to that growing element of student culture—intercollegiate athletic competition.

At Penn State there was very little in the way of sports before 1875—mandatory manual labor, lack of facilities, and isolation from other schools made athletic competition rare. From contemporary records,

Sis! Boom! Ah! Cuckoo! Pennsylvania State!
Yell! Yell! Yell! Again!
We're from the Land of William Penn.
State! State! State!
Whiskiwah–wah! Biskiwah–wah!
Holy Moki!! Pennsylvani!! State!!

FIGURE 3 A football souvenir card of the early 1900s includes a college cheer, pennant, and seal

FIGURE 4 1877 baseball team

FIGURE 5 Baseball game on the original Old Beaver Field location, ca. 1877–78. Window frame in upper-left corner shows that photo was taken from an upper floor of the Chemistry-Physics Building

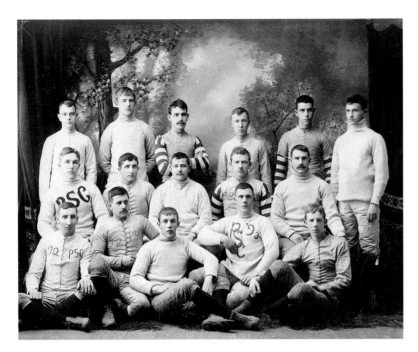

FIGURE 6 1890 football team

FIGURE 7 Old Engineering Building, ca. 1900

we know that cricket was occasionally played on the Old Main lawn in the 1860s by students from the Philadelphia area. Students began playing baseball on campus and against clubs from local communities as early as 1866, when the team traveled to Lock Haven for a game. By 1871 the faculty allowed students to use as much of the college grounds as needed for baseball. Four years later, the faculty granted use of ground to the northeast of Old Main for baseball and other athletic purposes. By the end of the 1870s, regulations were created on the use of the front lawn for sports, including times when sports were permitted there, and a rule requiring student-athletes to be responsible for the behavior of their "guests" (i.e., the visiting team).

In 1881 a group of students led by Irvin P. "Pit" McCreary challenged students at Bucknell to a game of "foot ball." In fact, the rule book used for this

contest survives in the University Archives, inscribed by McCreary with the details. The game they played was more akin to rugby, from which American football was developed in the 1880s, but the players and officials' names are listed inside the front cover, as well as the score, 9–0, Pennsylvania State over Bucknell.

Baseball was still the dominant sport on campus then—the first intercollegiate baseball game in 1882 also had Bucknell as the opponent—and tennis was popular as well. Football games with other colleges did not resume until 1887. Bucknell, our closest collegiate neighbor at the time, was again the competition (a proposed game with Dickinson earlier that fall failed to come off). Nevertheless, Penn State achieved its first undefeated season—beating Bucknell in two matches, the first played at Lewisburg and the second on campus in State College. A modest beginning, to say the least, but it was the start of an uninterrupted tradition of college football for Penn State.

A Formal Athletic Field Is Created

Football is a game of boundaries, and thus a demarcated field is required to know whether a player is "out of bounds" or has touched the ball down past the goal line to score. With the addition of points gained by kicking the ball between a pair of goalposts, and lines on the field to help in determining whether a team has advanced the ball far enough to receive an additional set of plays, the essential elements of the field are complete. But where to put it?

The Penn State campus in the 1880s still occupied the original 400 acres, half provided by donation and half by sale to the trustees of the Farmers' High School by Centre Furnace ironmaster James Irvin in 1855. Roughly bounded by today's College Avenue, Atherton Street, Park Avenue, and Shortlidge Road, the campus was expansive, but quite empty. Of the first structures, the College Building (later called Old Main, but initially just "the College") and the president's house are still in the same locations, although that original Old Main was torn down and rebuilt in 1929–30, and the president's house (much remodeled over the years) is now the front part of the Hintz Family Alumni Center.

Starting in 1887, the state legislature began appropriating funds for new buildings at Penn State. Half a dozen new academic structures were designed by architect Fred L. Olds in the modern architecture of the day, which incorporated elements of both the ornate Romanesque and Neo-Gothic Revival styles. The massive engineering building, completed in 1893 and later destroyed by fire in 1918, was the crowning

achievement of this building boom; it was the largest academic building on campus and, with 70 percent of the student body enrolled, exemplified the dominance of engineering at Penn State.

Olds also designed six faculty residences, wood-shingled Queen Anne–style structures scattered around the larger brick and stone buildings, creating a delightful, spacious campus, laid out on an almost rectangular grid of roads and paths.

In the middle of this academic village, almost assuming the role of town green, lay a space initially designated by the faculty in 1875 for athletic activities. By 1890 it included a semblance of a running track, baseball diamond, and a field for football, all usually in poor condition and sometimes completely unusable. In addition, there was a very crude grandstand of wood construction. It had a tiered arrangement of bench seating for perhaps 100 people, and a very light, forward-sloping shed roof for cover. Here students and other fans watched early track meets, and baseball and football games. But a better facility was needed.

In March 1891 the Trustees' Executive Committee voted to accept the request of the students' Athletic Association to formally designate the grounds they

FIGURE 8 Faculty cottages along Pollock Road, south of Old Beaver Field. Two of these houses (Spruce and Pine) were later moved to behind the Ritenour Building, where they still stand today

FIGURE 9 The unimproved field and simple grandstand, ca. 1890

FIGURE 10 Old Beaver Field grandstand, with bleachers on left; Ag Experiment Station and barns lie to the north of the field on "Ag Hill"

had been playing on for a number of years as the college's athletic field and authorized President Atherton to expend up to $150 to aid in grading the grounds.

The *Free Lance*, hoping to boost enthusiasm for additional athletic facilities (including a separate gymnasium that remained unfunded by the legislature), began to advocate in 1892 for seeking a wealthy donor to construct a field, which would naturally be named in the donor's honor. There were very few names to choose from, but one came to mind, Captain Charles W. Roberts of West Chester (the rank of captain coming from Civil War service). Roberts was a businessman and a member of the Penn State Board of Trustees and its building committee. He had funded military prizes for student achievements in the college's corps of cadets, and the place of their week of spring drill near Pleasant Gap was named "Camp Roberts" in recognition of his support. He had also contributed funds for furnishing the Ladies' Cottage (the women's residence hall) and one of the campus's new academic buildings completed in 1890. It is not known if Roberts was approached for a donation to help with athletic construction; ultimately, however, the naming decision went in a different direction.

General Beaver apparently sympathized with the students' desire for better facilities. It was not surprising, for Beaver had always been a supporter of the students and their interests, telling every freshman class at their orientation, "This is my class!" He sometimes intervened directly with faculty to argue for students' rights and even mediated a student strike against the faculty in 1906.

In Pennsylvania it was not common for public colleges to receive state funding for athletic purposes, which were usually financed by student fees and alumni gifts in most institutions. However, as governor, Beaver used his influence to add a line to Penn State's legislative appropriation for the 1891 biennium to provide $2,000, and a further $1,000 in 1893, for the improvement of the athletic grounds. These funds would enable the college to both lay out a quarter-mile track, including a baseball diamond and football field within the track and tennis courts nearby, and construct a larger grandstand. The oft-cited figure of $15,000 for this construction was apparently confused with the cost of moving the field in 1908; by comparison, $13,500 was appropriated for grading and maintenance of all campus grounds, roads, and walkways in 1887.

In recognition of General Beaver's vital role, the athletic association, with the trustees' concurrence, named the new field for the governor in June 1892. Old Beaver Field, as we call it now to differentiate it from its successors, was completed in 1893 on the site of today's parking lot behind Osmond Lab.

The new grandstand was basically a hip-roofed house without side walls. The roof was supported by a row of six columns along the front and rear, with an additional column at the midpoint along the sides. At the roof's center point, a gabled portion projected toward the front. Contemporary photographs show "Beaver Field" lettered along a curved line at the top of the gable face. Beneath it, "1893" appeared and below, on a second line, was "P.S.C."

There were individual backed seats in tiered rows that accommodated approximately 500 spectators, and immediately adjacent, but outside, was bleacher-type bench seating. Also designed by college architect F. L. Olds, the grandstand, called an "athletic house," included a room underneath with lockers, closets, and "a shower bath," supplied with water brought from the Experiment Station building, situated to the north on "Ag Hill," which became

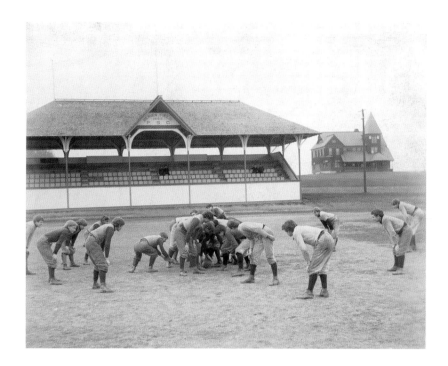

FIGURE II Football scrimmage on Old Beaver Field. Gabled front of grandstand shows the name and date of the facility

the setting of most of the college's agricultural work (including the location of Armsby, Patterson, and Weaver buildings) in the twentieth century.

Old Beaver Field would serve the college for sixteen years, hosting both football and baseball games as well as track and field events. Directly adjacent to the west end of the field, a three-story building was erected in 1904 and largely paid for by alumni donations. Dubbed the Track House, it served as an athletic dormitory and storehouse for outdoor sports gear and sat at about the location of today's Whitmore Lab. The main athletic facility sharing the center of campus with Beaver Field was the nearby college Armory, which was pressed into service as gymnasium and the venue for indoor "athletics" meets, which included a few track and field events, gymnastics, boxing, and even a three-legged race and a tug-of-war. The Armory was also later used for

FIGURE 12 Track meet at Old Beaver Field, ca. 1905

FIGURE 13 Baseball game on Old Beaver Field; looking southwest, Track House on right. Other structures (left to right) are George G. Pond residence (later Spruce Cottage), McAllister Hall, and Chemistry-Physics Building, with tower and northwestern corner of Old Main just visible behind Chem-Physics

FIGURE 14 College Armory, with members of the Cadet Corps at attention

basketball and wrestling. These events were usually interclass competitions, as they had originally been for football and baseball, but gradually intercollegiate contests would be added to the sporting life of the campus.

Engineering was becoming the dominant force on campus in those years, as Pennsylvania's industrialists were grasping the value of college work in quickly developing fields. With major business leaders like Andrew Carnegie and Charles Schwab serving on the Board of Trustees, Penn State was insuring its future as a college with strengths in the sciences and technology. It was still a relatively small and isolated institution, but it would grow rapidly in the next two decades as its students developed their own extracurricular activities for both entertainment and the development of the character and skills that would serve them well in the new twentieth-century America. Gradually, participation and interest in athletics played an ever more prominent role in the life of the student body. As a result, these activities would strengthen the public's awareness of Penn State and the college's growth would soon impact the location of Beaver Field on campus.

Dunbar Rabenau

Kiser Murray Davis Taylor

Cheer Leaders 1923-24.

"WISH-WACK, PINK AND BLACK, GO STATE GO"

In 1869 Princeton and Rutgers played the first intercollegiate football game. This was the official beginning of American football by all accounts, but not of intercollegiate sports. A rowing match between Harvard and Yale in 1852 at a New Hampshire lake is credited as the first intercollegiate sporting event, which soon led to other competitions. Track and field events were common, and baseball grew in popularity during the Civil War, and soon after at many colleges.

The long-established eastern schools, Harvard, Yale, and Princeton, quickly became the dominant powers in college football and in the Intercollegiate Football Association, a league formed in 1876. The "Big Three" regularly played before tremendous crowds—for example, the 1887 Harvard-Yale game had 23,000 spectators in attendance.

Penn State joined the ranks of schools playing intercollegiate football in 1887, eighteen years after that first Princeton-Rutgers game. At that point, in 1869, Penn State was struggling to survive as an institution. It had just 45 students in residence, 12 faculty members, and the only buildings on campus were Old Main, the president's residence, and the college barns. By comparison, Yale's enrollment in 1869 was 736 and by 1887 it had 1,245 students; Penn State had grown to only 167 students by 1887.

When Penn State began playing on Old Beaver Field six years later, its enrollment totaled 317 students; by 1908, the last season for Penn State football games on Old Beaver Field, there were 1,291 students—a fourfold increase in those sixteen years. During those years, Schwab Auditorium, Carnegie Library, and McAllister Hall were built near Old

FIGURE 15 Schwab Auditorium, 1902

FIGURE 16 Armsby Building (called "Main Agriculture" when built in 1905)

Main, and farther northeast of the athletic fields new structures for agriculture, including Armsby, Patterson, and Weaver buildings, joined the barns on Ag Hill. Faculty numbers increased to 117, and the college's curricula were organized into seven schools with more than a dozen courses of study.

Penn State Football Becomes an Intercollegiate Sport

As at all colleges, Penn State football was the product of student enthusiasm for playing sports and, soon, for athletic competition against other colleges. Land-locked Penn State never competed in rowing in these early years, and so baseball pioneered and was very popular. But football soon caught the students' attention. From pickup games and interclass contests, students quickly developed interest in having a college "eleven" to test their mettle against teams from other schools.

In the 1886–87 academic year, the students formed an Athletic Association to organize support for their teams. Its first president was Gilbert A. Beaver, General Beaver's nephew. The association raised money to equip and outfit teams as well as for travel expenses to play other schools. It was a difficult job at first, however, with inconsistent numbers of students involved and paltry finances. Initially there were no admission charges to home games. Penn State's early away games rarely brought in more than $30 to $40 for each outing, barely enough to cover travel expenses, and guarantees paid to cover visiting teams' expenses for Penn State's few home games usually came from contributions by local merchants. Most funds for athletics came from association member subscriptions, player contributions, and the proceeds of benefit concerts, debates, and dances. In 1891 students voted to levy a $1 assessment fee to support the hiring of a director of physical training for students, who also developed conditioning exercises for football players. By 1896 increased annual fees provided $2 for football and $1 for baseball per student and, in 1899, the Athletic Association was permitted to collect a $5 assessment fee from each student to boost its modest funding to support athletics.

While the association had a committee to oversee each sport, the team itself, in particular the captain and the manager, played the central roles in these early years. The captain selected the players who would participate and guided conditioning, practices, and the play of the game. The manager's role was almost as important; he corresponded with athletic associations at other schools to schedule games, negotiated financial arrangements, and kept track of the team's funds. In 1891 the association helped create a short-lived Pennsylvania Intercollegiate Football League of six schools, including Penn State, Bucknell, and Dickinson. Penn State was named the league champion for the 1891 season, its most significant achievement in football to that point. The league disbanded the following year, but Penn State's record for the two seasons was an outstanding 11–3.

During the 1890s football gradually changed; game management responsibilities moved from team captains to coaches. Penn State did not have a football coach until 1892. George W. Hoskins is usually credited with being Penn State's first football coach, and the early coaches were hired at the urging of the student body. They were primarily expected to develop physical education work for male students as well as a training regimen for athletes. The "coach" might even occasionally play in a game if the team was short a man. William N. "Pop" Golden

was probably the first "head coach" who advised the students on the play of the game, an early form of a coach's main duty today.

While Golden is identified in official records as head coach from 1900 to 1902, his role as director of physical training and then director of athletics, until his departure from the college in 1912, came to be more significant. He also served as an informal coach for track and, later, basketball and also helped plan and oversee the development of Penn State's new athletic facilities in 1908.

The coaches who succeeded Golden—Dan Reed (1903) and Thomas Fennell (1904–8)—were still part-timers (Fennell even maintained a law practice in Chemung County, New York, during his coaching tenure). Gradually, the role of physical training directors and coaches began to grow as they introduced game strategies and plays to the teams as well as carrying out their traditional roles of training and physical preparation for the games.

The control of finances, too, was changing, gradually moving from the student manager to alumni. The roles of the college president and faculty were limited to granting permission to play off-campus and to miss classes for athletic activities. The Board of Trustees dealt with fees and scholarships, the hiring and contracts of athletic trainers (football was not the only sport with a designated trainer), and seeking state appropriations for athletic facilities, including a gymnasium, as part of their responsibility for developing the college's physical plant.

As the student Athletic Association began to continually lose money, alumni and faculty stepped in. The college could not afford to be responsible for student organization debts, so graduate managers were appointed to oversee finances for athletics as well as for the Penn State Thespians and the *State Collegian*

FIGURE 17 Penn State vs. Penn in Philadelphia, 1900, a 17–5 loss

Alumni Advisory Committee requested financial aid and the Board of Trustees voted in 1901 to provide ten scholarships for athletes, probably the first college-sanctioned athletic scholarships in America. Starting in 1901, the team had seven straight winning seasons.

Although the number of games played at Old Beaver Field gradually increased to between four and six per season between 1902 and 1908, visiting State College was not an easy or inexpensive trip. With a total seating capacity of 500, Penn State could still only attract other small Pennsylvania colleges, such as Dickinson, Gettysburg, and Susquehanna, as well as noncollegiate opponents like the Jersey Shore Athletic Club, Dickinson Seminary, and the Bellefonte Academy for home games. In those years West Virginia was the only college from outside the Commonwealth willing to travel to State College for a game, and Pitt (then the Western University of Pennsylvania) played Penn State twice in Bellefonte in 1900 and 1901, rather than coming all the way to campus.

Thus, most games were played away from campus, some at neutral fields in cities where larger crowds could draw more ticket sales revenue, such as Williamsport, where 8,000 fans watched Penn State defeat Dickinson in 1905. After the turn of the century, Penn State began to play more top teams— Penn, Princeton, Yale, Army, Navy, and Pitt—in away games with guarantees totaling about $3,000 a season. In fact, Penn State's share of the receipts for just the Thanksgiving Day games in Pittsburgh accounted for about half of the football team budget in the 1904–6 seasons. At the same time, home games, which accounted for less than half of the annual schedule, only brought in a total of about $100 in a typical season.

newspaper. The managers were alumni (hence the adjective "graduate"), and soon they became college employees and also took on the administration of the Alumni Association. This demonstrated the growing role of graduates in the life of the college, and they frequently attended college sporting events and special campus weekends like homecoming, which was first held in 1920.

While the early 1890s were a time of relative poverty for Penn State's Athletic Association, the quality of Penn State football was improving between 1891 and 1894, and Penn State had its first All-American, Carlton A. "Brute" Randolph, named to the third team by Walter Camp in 1898. By the late 1890s the numbers of games played in a season increased to nine or ten, but Penn State was still only playing two or three games at home each year—about a quarter of all its games. After a stretch of losing seasons, the

How the Game Was Played

In the three-and-a-half decades between the first college football game and the establishment of the NCAA in 1905, much about the game was formalized and consistent rules were agreed upon. By the time Penn State started playing football on a continuing basis in 1887, team sizes had been set at eleven men and the ball used in play had changed from the rounder ball used in the early games, which resembled today's soccer ball, to an oval ball that came from the rugby tradition. The means of scoring and points awarded for each score were still changing, eventually giving more scoring prominence to running the ball than kicking it.

The earliest games of football at Penn State, both informal and organized, were played on the lawns around Old Main and then on the fields to the northeast, which in 1893 became the home of Old Beaver Field. Beginning in 1876 the field size was set at 110 × 53 1/3 yards (in 1912 it was enlarged to 120 × 53 1/3 yards, including two ten-yard end zones, which are still the dimensions today). The goalposts marking the goal line for scoring points by running or kicking had taken on the classic "H" form. Yardage lines were added once the system of "downs" for retaining possession of the ball became a part of the game, starting in 1882.

The most controversial aspect of the game in these years was the use of "mass momentum" or "wedge" plays, the "flying wedge" being the most remembered today. These set plays featured most of the players starting from the backfield and running closely alongside the ball-carrier to overpower those attempting to stop the advance. The collisions these plays produced were quite dangerous to players, who wore little padding or protection, and led to many injuries.

However, the injuries were far overshadowed by the periodic fatalities from football games at all levels. With nineteen deaths in 1905 alone, several in college games, a national call for change led by President Theodore Roosevelt resulted in a White House meeting with leading college officials on October 9, 1905. Gradually, reforms in running plays and tackling emerged from the ongoing discussions. The dangerous flying wedge play had been banned in 1894, but other mass-momentum plays were made obsolete by both new rules and other types of plays, such as the forward pass.

From all the reforms finally came the realization that interinstitutional controls enforced by a formal organization were needed, and in 1905, following the president's White House meeting, came the establishment of what would soon be called the National Collegiate Athletic Association, the NCAA, to fill that need. However, despite formalizing the process of change and standardizing the game between 1880 and 1905, one major area of contention remained an issue for the following forty years.

The eligibility rules for athletes to compete were still an inconsistently handled matter in almost all

FIGURE 18 Penn State vs. Allegheny College, 1904, on Old Beaver Field in one of four home games that season. Penn State emerged victorious by a score of 50–0

colleges. Of particular concern were the issues of allowing students to play who did not meet academic standards, and of "tramp athletes" who transferred from one school to another with undue frequency in order to play a sport. But the most disputed problem turned out to be the eligibility of athletes who accepted cash to play on summer baseball teams, which helped cover their college expenses. This seemed to strike at the heart of amateurism in college athletics and it was the same issue that later cost Jim Thorpe his medals from the Stockholm Olympic Games of 1912.

How the Fans Enjoyed the Game

It is remarkable how quickly football, particularly the college variety, became a national craze. In 1880 a relative handful of colleges were playing football; twenty years later, Penn State was one of perhaps 100 schools fielding a team. Football was originally a northeastern sport, but as early as 1879 the University of Michigan played an intercollegiate game against Racine College of Wisconsin, probably the first college game played west of Pennsylvania. Within sixteen years, the predecessor of the Big Ten conference had been founded (1895), and football was spreading to the West Coast and into the Deep South. Penn State, in only its seventh season, bravely played the "southern champion" University of Virginia in 1893, and won by a close score of 6–0.

The "Big 3"—Harvard, Yale, and Princeton, along with Columbia—had begun the Intercollegiate Football Association in 1876 and were soon playing their championship game at New York's Polo Grounds. There were 15,000 spectators for the 1883 game, and crowds of 40,000 or more were typical

in the 1890s. Faculties at these schools expressed concern about the amount of time students spent on games in big cities and away from the classroom. In partial response, the schools built massive on-campus stadiums. Penn's Franklin Field was begun as a track facility in 1895 but was "built out" for football and other sports in 1905, eventually seating a maximum of 78,000 fans. Harvard erected its stadium, which held almost 35,000, in 1903 (expanding to almost 58,000 in 1929). It was followed by Princeton's Palmer Stadium, which accommodated 46,000 fans, in 1914, and in that same year the 71,000-seat Yale Bowl was opened—the first "bowl" in history to exceed the size of the Colosseum in Rome.

The first championship game, in 1876, pitting Yale versus Princeton, was held at the Hoboken, New Jersey, cricket grounds and drew an enormous crowd. Large contingents of the schools' student bodies yelled back and forth with the Yale ("Rah-rah-rah-rah") and Princeton ("Hurrah-hurrah-sis-boom-ah") cheers. Also attending was a fashionably dressed crowd from across the Hudson, watching in carriages from the edges of the cricket field and displaying flags and blankets in the schools' colors. The 1893 game, in New York City, between the same opponents was preceded by a four-hour parade up Fifth Avenue.

Obviously, Penn State could not match this opulence and spectacle, yet the games at Old Beaver Field and other venues were no less hard-fought and cheered by the fans. By the time Penn State played its last season at Old Beaver Field in 1908, most of the basic elements of the classic football experience were in place.

In those earliest games against Bucknell, Penn State had new, matching uniforms of pants and jerseys in the school's colors. In those 1887 contests, the college yell—

Yah! Yah! (*pause*)
Yah! Yah! Yah!
Wish-Wack! Pink-black!
P! S! C!

—urged on the boys in their pink and black uniforms. They were the colors of the senior class when they were adopted by the Athletic Association for the college teams. It is often said that three years later the association voted to switch to dark blue and white because the pink clothing had faded from the sun or too many washings. Another possibility, as reported in the *Free Lance*, was that pink was abandoned because the proper shade of cerise could not be ordered for additional uniforms. Yells changed as a result, but the cheers were no less enthusiastic:

Sis! Boom! Ah! Coo! Penn State!
Yell! Yell! Yell! Again!
We're from the land of William Penn!
State! State! State!

The faculty permitted students to miss classes to attend one away game in the 1890s—80 students traveled by train to Lewisburg to join the 2,000 fans for the 1893 Bucknell game. After 1899 the privilege was expanded to two games, although both had to be played within the borders of Pennsylvania. It was a great treat whenever students were permitted to travel with the team—all types of "pranks, gags, and 'hi-jinks'" were customary, and betting on the games was common among students, especially against the big rivals of the 1890s, Bucknell and Dickinson. Clearly, away games, notably against Penn in Philadelphia, and the new tradition (starting in 1904) of a Thanksgiving Day game in Pittsburgh against Pitt, became very attractive to alumni in and near those two cities.

All manner of pennants to wave, souvenir cards to display, and rosettes and pins to wear became popular. Athletes proudly showed off their blue varsity "S" letters on white sweaters beginning in 1894. Football players on campus, even the "scrubs" who didn't see much playing time, were honored and students routinely watched practices on Old Beaver Field. Flamboyant newspaper stories portrayed students and fans stylishly dressed and showing "Blue and White fluttering through the crowd," creating visions of spirit and pride for alumni who could only read about the games.

The evolution of a sporting press parallels and complements the development of sports in America. Some press coverage of horse racing and boxing could be found as early as the 1830s. After the Civil War and the emergence of baseball, magazines and, increasingly, newspapers began to report the results of games and thereby promote interest in sports as a legitimate pastime and symbol of American culture. In the 1880s sports pages became staples of American newspapers; as public interest in sports grew, the larger readership attracted more advertising revenue, reinforcing the importance of sports coverage in journalism.

Penn State's games were covered initially by local press; the 1887 games with Bucknell had a detailed write-up in the *Free Lance*, unlike the 1881 game (which was only briefly mentioned in Bellefonte's *Democratic Watchman* and the Bucknell student paper, as there was as yet no Penn State student newspaper). But very quickly, the popularity of college football meant that Penn State's games were reported in vivid detail by both large and small papers across the state. Although it was rare for a city reporter to venture to State College in those days, Penn State games played in Philadelphia, Pittsburgh, and Williamsport all drew extensive coverage of the "boys

FIGURE 19 A souvenir card showing a coed waving her pennant, 1903

[It is with much reluctance that we accept the resignation of Mr. Morris. He has been an agreeable and energetic co-worker, and has managed most efficiently the business affairs of THE FREE LANCE, as is fully shown by the above statement.—EDITOR.]

FOOT-BALL.

P. S. C., 54—BUCKNELL, 0.

The regular practice and careful training of our "first eleven" developed confidence enough in the playing abilities of our boys to lead many people to believe an interesting and exciting game would be played on the occasion of our visit to Lewisburg Nov. 5 ; and the State College foot-ball team did not disappoint the admirers at home nor did they leave the opponents' field until they had gained admirers there. It is a fact worth mentioning, that our boys made their first "touch down" within two minutes after the game had been called.

The playing on both sides was very commendable throughout the entire game.

The teams were well matched in size and strength. Neither side could gain anything by what is called rough playing, and all the points scored were made by the skillful playing of "tricks," the best of which was the one so frequently played by our half-backs.

The game ended with the score : Pennsylvania State College, 54; Bucknell, 0.

BUCKNELL, 0.—P. S. C., 24.

The Bucknell University foot-ball team, of Lewisburg, Pa., which was defeated November 12th by the State College eleven by a score of 54 to 0, played a return game here November 19, and again lost by 24 points to 0, or two goals from touch-downs and two safety touch-downs to 0. Hanson and Shipman did the best playing for the visitors, while the honors for the College eleven were divided between J. P. Jackson, Linsz (captain) and Barclay. The kick-off was made at 10 A. M., Captain Hanson, of the visiting team, winning the toss. During the first half the playing was very close and exciting, only one touch-down being made, and that by Jackson. In the second half the College played a strong rushing game, excelling their opponents in running, tackling and kicking. Captain Linsz secured three touch-downs from two of which Mitchell kicked goals. The running and rushing of Linsz, and the brilliant tackling of Barclay were the features of this half. The teams were as follows :

BUCKNELL.		P. S. C.
Hanson	Quarter Back	Linsz
Catterall	Half-Backs.	J. P. Jackson.
Campbell		J. G. Mitchell.
Shipman	Full-Back	Mock.
Tustin		Leyden
West		Rose.
Booth		Hildebrand.
Rhines	Rushers.	Kessler.
Farrow		Weller.
Kirkendall		McLean.
Wolfe		Barclay

N. E. Cleaver, of the Pennsylvania State College, was referee, and J. S. Braker, of Bucknell University, umpire.

LOCALS.

—Wanted—a match—Horace.

——Electric lights have been placed along the path.

—Prof. Jackson was on the sick list a few days this week.

—The college walks are being covered with broken shale.

—The Juniors are kicking on account of too many examinations.

—The new cadets know what every stone for miles around looks like.

—Tennis is still played at P. S. C., although it is the 15th of December.

—Twenty buildings have been erected at State College during the past year.

FIGURE 20 *Free Lance* account of 1887's two football games against Bucknell

from up in the mountains," especially if they defeated an urban school's eleven.

The college yearbook, *La Vie*, included an extensive review of each football season. In the 1908 edition this included write-ups of the Carlisle Indian School game at Williamsport from the *Pittsburg Press*, the Navy game at Annapolis from the *Philadelphia Press*, and the *State Collegian*'s account of the Western University of Pennsylvania game on Thanksgiving Day in Pittsburgh. The last story began with a verse celebrating the 6–0 victory and the exploits of Penn State's earliest first-team All-American player, center W. T. "Mother" Dunn.

> When the State man puffs his trusty pipe,
> Oh, many years from now!
> And his happy child looks up and asks
> The where, and why, and how;
> This story will again be told
> Of our heroes tried and true,
> Who fought with dear old "Mother" Dunn,
> 'Neath our glorious White and Blue!

> Battered and bruised by the hardest and most trying gridiron campaign that any State eleven has ever experienced, our invincible Varsity went to Pittsburg for the final battle, and achieved a Thanksgiving Day triumph that put a glorious finish to the most successful season in our gridiron history.

Even with local and statewide press coverage, news of away games was eagerly awaited by the student body. The use of the telegraph to speed news across the country not only helped newspapers to report game results more quickly, but also enabled students remaining on campus to closely follow the progress of these contests by telegraphed reports through the course of the game. Starting in 1910, these telegraphic reports came from "Marconi Wireless" stations, whose transmissions of the details were picked up by the college's experimental radio station, run by the Electrical Engineering faculty. Whenever a play was relayed to the crowd at the auditorium, a cutout image of a football placed on a large board replicating the gridiron was moved to show the action in progress.

Another indispensable part of a football game was the music of the college band. Originally, armies created bands as aids to marching and signaling battle maneuvers. As football is often likened to combat and warfare, so the marching band transitioned almost seamlessly into college sports and especially to football. Penn State's Cadet Bugle Corps, consisting of six musicians, was organized in the fall of 1899 and led by Spanish-American War veteran and bugler George H. Deike '03. Among the student body, there were twenty-five musicians, but only half owned instruments. After the Board of Trustees turned down a request for funds for the band in 1900, trustee Andrew Carnegie donated $800 to purchase a variety of drums and brass instruments. The newly expanded band played at numerous events, including several fundraising dances for the football team in 1901.

More important, the band played college fight songs at both 1901 home games (against Susquehanna and Dickinson) and for the Lehigh game at Williamsport, traveling there by train along with 200 other students and fans. In 1902 the band led the parade of Penn State supporters from dinner at Carlisle's Hotel Washington to the field for the Dickinson game. In 1903 the band played at the Washington & Jefferson College game in Pittsburgh, a Penn State victory over what was then one of the most powerful teams in the state.

Increasingly, the competition between college bands to inspire the spirit of team and fans became part of the tradition of Penn State football games. Indeed, the Cadet Band's primary role soon became to support athletic contests—football and baseball games and even track meets. However, like all extracurricular activities of the day, the band was "student-led and student-organized"; it supported itself through concert performances and other events, receiving no college funding or supervision.

We may not know if the band performed any type of halftime show in this era, but we can safely say that there was no Nittany Lion mascot roaming the sidelines to raise the spirits of the crowd. It was only in 1907 that the Lion become the generally accepted symbol of Penn State, and the sport that gave birth to the Lion mascot was baseball, not football. It was Harrison D. "Joe" Mason '07, third-baseman and captain of the college nine, who came up with the impromptu story of the prowess of the Nittany mountain lion when shown the Bengal tiger mascot on a road trip to Princeton. A story advocating the idea of a Nittany Lion mascot first saw print shortly after in *The Lemon*, the college humor and satire magazine of the day, which happened to be edited by the very same Joe Mason.

But there appeared to be some confusion from the start over genus and species: the 1907 *La Vie* featured a poem dedicated to "Old Nittany," retelling the story of the lion's roar inspiring the team to defeat the Carlisle Indian School 4–0 in 1906 before 5,000 fans in a driving rain at Williamsport. It concludes:

> Now may the lion ever stand
> The mascot of our favored band,
> The hope of all that's true and great.
> All hail the "King of Pennsy State."

FIGURE 21 "Joe" Mason, varsity baseball player and editor of *The Lemon*

FIGURE 22 Football coach Hugo Bezdek with
the Lion mascot, still in his African lion guise,
in the early 1920s

FIGURE 23 Cheerleading squad, 1923–24

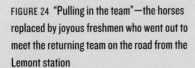

FIGURE 24 "Pulling in the team"—the horses
replaced by joyous freshmen who went out to
meet the returning team on the road from the
Lemont station

Ironically, the poem is followed by a picture entitled "Old Nittany," but showing an African lion. Still, there was no "man in the suit" until Dick Hoffman borrowed the lion suit he wore in a Penn State Players' production of *Androcles and the Lion* to wear on the sidelines of the 1921 Georgia Tech game, won by Penn State 28–7 before 30,000 fans at the Polo Grounds in New York. Hoffman continued donning the lion suit for games until he graduated in 1923, and the lion later appeared briefly in the 1927 season but was banished by Coach Bezdek for bringing bad luck to the team. The mascot didn't appear again until 1939, and then in the guise of a mountain lion—the tradition we still honor.

While cheers and yells had been a part of college traditions from before the time of intercollegiate sports contests, every class had its own yells, along with other examples of school spirit, including the college song, the "alma mater," with lyrics written in 1901 by Professor Fred Lewis Pattee to the tune of one of his favorite hymns, "Lead Me On."

In addition to class yells, the college cheers were taught to incoming freshmen by upperclassmen at mass meetings in the fall. Soon one of these men became a designated "Yell Leader" at football games—the first cheerleader. In the early 1900s he was joined by an assistant, designated the "Song Leader." Songs such as "Victory" and "The Nittany Lion," written by Jimmy Leyden '14, were sung at every game (and are still played by the Blue Band today) and their lyrics were memorized by new freshmen as a part of student customs.

Soon, the two leaders were joined by a squad of men under their command, who with megaphones, white pants, and sweaters with a blue "S" became the earliest cheerleading squads. However spirited and boisterous the student and alumni cheers

became, these young men stood their ground; there were no acrobatics or performances of the kind we have come to expect these days, and, as in the band, there were no women on the field helping to lead those cheers.

Students, both men and women, attended mass meetings and rallies preceding both home and away games and, even in defeat, returning teams were feted with "welcome back" parties out of loyalty to the school. A victory, or a tie against a "Big 3" team, was gloriously celebrated. If the team returned by the Pennsylvania Railroad to the Lemont station, built in 1885, the entire student body would go out to meet the train with all the available carriages. The students would take the place of the horses and "pull the team in" the three miles to the campus.

A great victory would be celebrated with a massive bonfire on campus, usually held on the open field to the west of the Armory (about the location of today's Electrical Engineering buildings). Freshmen would be tasked with retrieving "surplus" wood from town and campus—any loose planks, fence posts, or pieces of outhouses and small sheds would do—and building the bonfire. The team, president and faculty, student body, and townspeople would gather for the ceremonial lighting and enjoy the speeches, cheers, and music accompanied by the blaze.

Despite an occasional mishap—a 1914 bonfire celebrating a tie with Harvard turned into a near-disaster when students mistakenly doused the massive wood pile with gasoline, rather than the usual kerosene, resulting in an explosion that broke hundreds of windows and put several students, including quarterback Elgie "Yegg" Tobin, in the hospital—bonfires remained part of both postgame celebrations and pregame pep rallies for many years.

Bonfire After Cornell Victory
State 29. Cornell 6.
Oct 19, 1912.

FIGURE 25 Bonfire preparations, celebrating 13–13 tie with Harvard, 1914

FIGURE 26 Bonfire, 1912

Football Provides Public Recognition for the College

While the Allegheny Athletic Association and the Latrobe Athletic Association fielded professional teams in 1896 and 1897, a multistate league of professional football teams did not become a reality until 1920. Football in these years was for most the quintessential college game. It also quickly became a game driven by the desire to win and the prestige it would bring. While Harvard's and Yale's initial models for athletics were Oxford and Cambridge, where gentlemen competed with dignity and sportsmanship for the honor of victory, America was without a hereditary aristocracy and believed it had little tradition of class distinctions. It was more democratic and valued the story of the honest, hardworking young man who could rise above a hardscrabble background to become a success. In America, college sports were not simply a pastime for the elite as they were in England, but an ideal unto themselves. Accomplishment in athletics enabled young men to express both physical prowess and moral rectitude.

Just as important, the glory and honor associated with a winning team and outstanding players conferred a status and distinction upon a college that would have been hard to achieve in the popular mind through other means. While today Ivy League schools like Harvard, Yale, Princeton, and Penn are in the top ranks of America's most distinguished universities for their faculty, academic programs, and highly selective student bodies, in the nineteenth century they were the elite of American athletics. For years, the presumed champions and many All-Americans of college football came largely from these four schools. They regularly played in massive stadiums and in

Thanksgiving Day games to crowds of as many as 70,000 fans.

Unlike the dominance of Oxford and Cambridge in England, America had a vast collection of colleges and universities of various types. Academic programs and "star" faculty did not yet provide meaningful distinctions to the public, and it was often athletics that generated regard for a particular college or university. For an isolated small college in the mountains, still a "cow college" in the minds of many in Pennsylvania's cities, winning football teams, outstanding athletes, and other sports successes gave Penn State a public face for many who otherwise knew little about the school.

It is not without reason that we acknowledge the Atherton period at Penn State as fostering the beginnings of the modern university that we know today. In both academics and the extracurriculum, it all came together in the years of 1882 to 1906, which mark George W. Atherton's presidency of Penn State and the creation of Old Beaver Field. The achievements of this period created the opportunity and need for expanded athletic facilities that marked a new era for Penn State.

NEW BEAVER FIELD
THE WOODEN ERA

The twenty-four-year tenure of President George W. Atherton transformed Penn State and created a firm foundation for a modern university. After such a long presidential term, it is not surprising that Penn State experienced a lull following Atherton's death; however, with the appointment of Edwin Erle Sparks in 1908, the situation soon changed.

The Sparks presidency (1908–20) was marked by the formal creation of extension programs and an expanded outreach to the state for support. Dr. Sparks had been a renowned scholar and one of the top extension teachers at the University of Chicago, and he put his enthusiastic personality and considerable public speaking skills to use in communicating Penn State's goals and needs to audiences all over the state. Even though funding remained sparse throughout his tenure, the college's academic curricula continued to grow and the number of undergraduates surpassed 3,200 by 1920.

The evolution of Penn State's football program paralleled the changes in the college. Between 1900 and 1906 Penn State recorded six out of seven winning seasons thanks to increasingly professional coaches who recruited and developed outstanding players, such as Penn State's earliest first-team All-American, center W. T. "Mother" Dunn, in 1906. The Penn State schedule expanded from five or six games to nine or ten per season, with the proportion of home games increasing from a quarter to almost half of all games in a season. Overall Penn State began to travel farther afield to compete with some of America's top-tier teams, such as Princeton, Harvard, Penn, Cornell, Notre Dame, and the Carlisle Indian School. Unfortunately, that success also hit a lull in

FIGURE 27 President Edwin Erle Sparks

increased by 50 percent, and this growth began to test the capacity of the college's academic buildings.

Paralleling the need for campus enhancements at Penn State was an increasingly recognized national movement for more rationalized urban designs. College campuses were not immune to city planning ideas inspired by the Chicago World's Fair of 1893, and symmetrical, neoclassical campuses and buildings were on the drawing boards, including large campus stadiums. The prescription for campus buildings seemed to be "if it's old, tear it down and build new; if you can't afford that, plant ivy to cover it up."

Penn State could not afford to demolish old structures—its building capacity for instruction and student life was already insufficient. Penn State's leaders concluded they needed a plan to both expand the number of buildings and create a more formal and functional layout. The planning effort likely began before President Atherton's death, continuing into the interim period when Governor Beaver served as acting president. The new plan became President Sparks's to carry out, and he enthusiastically proclaimed that a larger and more modern Penn State would even further awaken the people of Pennsylvania to the value of their state college for the Commonwealth.

Penn State's inaugural campus plan was created by New York landscape architect Charles Lowrie and approved by the Board of Trustees in June 1907. While the master plan was created by Lowrie's firm, deans and administrators contributed the details of the size and scope of new buildings and how they would serve modern needs.

Athletics and physical education were important parts of the plan. Moving the sports fields from the center of campus to its northwest corner to make way for new classroom and laboratory buildings was first to be implemented. "Pop" Golden, director of

the time of transition between Atherton and Sparks. Records went from 8–1–1 in 1906—the "finest season to date," according to football historian Ridge Riley—to 6–4 in 1907 and 5–5 in 1908.

Planning for the Campus and Sports Facilities

Although the nation's college-going population was still relatively small, the impact of college sports, particularly football, seemed to be drawing more popular attention to the excitement of attending college as well as to the value of a college degree. Penn State's physical plant had dramatically expanded in the Atherton years, but student housing was still woefully inadequate. Between 1906 and 1908, enrollments

athletics and physical training, was responsible for specifying the needs in these areas. In May 1907 he had already requested ground from the Board of Trustees for a second athletic field to supplement the outdoor sports and physical education activities that took place on Old Beaver Field. The fact that the field also hosted a variety of other large outdoor events, such as various freshmen-sophomore class scraps and military drill reviews, only reinforced the urgency. The trustees initially refused, but as they saw the need for a larger plan, they authorized him to begin a survey for a new Beaver Field in July 1907.

As athletic director, Golden advocated for the creation of "the best equipped most complete athletic plant in the college world" as part of the Lowrie plan. He was eyeing the northwest part of campus and believed he would need eighteen acres for the new fields and structures for his "great athletic park." He recommended moving the three-year-old Track House to the new area, building a similar building for visiting teams, and constructing a gymnasium, an outdoor swimming pool, and a "batting cage" structure with a glass ceiling and an indoor banked dirt track to enable both baseball and track teams to train during the winter months.

The centerpiece was to be the new football field and running track. Integral to its relocation was the decision to also move the existing grandstand, rather than to build a permanent stadium structure as was being done at wealthier schools. New academic and student buildings had a higher priority for funds. Since the existing grandstand was made of wood, it could easily be dismantled, reassembled, and later expanded as needed with additional wood construction.

The location of this sports complex was in a cleared, treeless space between two woodlots on the north edge of campus. The plan noted it was "high,

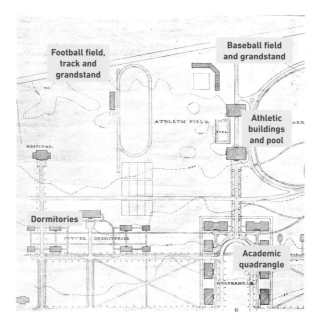

FIGURE 28 Detail of athletic fields, northwest corner of campus, in the 1907 Lowrie plan. Proposed athletic plant includes football field, track, and grandstand; baseball field and grandstand; gymnasium, baseball cage, and pool on center right; and tennis courts directly south of the running track. Hospital is center left, fraternities and men's dormitories in lower left, central academic quadrangle in lower right

FIGURE 29 Cider scrap—a part of freshman customs where freshmen and sophomores battle for control of a barrel of apple cider— on Old Beaver Field, ca. 1900

dry, and cool, yet protected on either side from the wind." On the east side of the field lay "Hort Woods," a large expanse of trees that extended from Park Avenue to today's Curtin Road. It gained its familiar name from the adjacent horticultural greenhouses and flower gardens located on its eastern edge. The boundary between the new athletic fields and Hort Woods was an unpaved road that became Allen Street, extending on into the College Heights section of the borough, which was developed starting in

1915. The woodlot on the western side would eventually become the home of the Nittany Lion Inn and Recreation Hall, with the college golf course beyond.

Both a gymnasium—which would permit physical education classes, basketball, and wrestling to be moved out of the Armory—and a batting cage to be used for indoor athletic practice sessions had been on the trustees' wish list almost since the Armory was built in 1888–89. However, the legislature was reluctant to appropriate funds for them. Now with plans drawn up, General Beaver predicted the General Assembly would approve the funding, and college officials began to approach alumni for donations to supplement the expected appropriation. Unfortunately, Penn State officials could not persuade the legislature. Golden then took matters into his own hands. He went to Harrisburg in the fall of 1908 and secured $15,000 for constructing the field and running track, and for moving the grandstand from Old to New Beaver Field. The rest of the athletic structures he had proposed, however, remained unfunded and unbuilt.

New Beaver Field

Between the 1908 and 1909 seasons, the covered grandstand at Old Beaver Field was moved to the west side of the new football field and track. The field was located in the space occupied by today's Nittany Parking Deck, Keller, and Mateer buildings. The date "1909" was added to the original date of 1893 on the lettering on the gabled façade of the grandstand. The field was dedicated on May 7, 1909, with the staging of Penn State's first interscholastic track meet for Pennsylvania high school teams. From this time forward, the earlier athletic field space would be identified as "Old Beaver Field" to distinguish it from New Beaver Field, and it

continued to be used for athletic practice, interclass games, and scraps. Significantly, the name of the new field continued to honor Governor Beaver, who had secured the state funding for the original athletic field and who loyally served the college as president of the Board of Trustees until his death in 1914.

The move to the northwest corner of campus brought the original 500 seats to New Beaver Field, and, shortly afterward, the field's seating capacity was expanded to more than 1,000 with additional wooden bleachers on either side of the covered grandstand. Before 1920 the west-side bleachers were lengthened and wooden bleacher seating was added to the east side, increasing capacity to nearly 6,000.

Penn State's teams achieved tremendous success with a 70 percent winning record (40–10–7) in the 1919–24 seasons. As a result, the west-side stands were reconfigured in the early 1920s. The original enclosed grandstand from Old Beaver Field was demolished, although the wood was recycled for other uses. This enabled a taller set of wooden bleachers to be constructed, extending farther along the sidelines with a larger numbers of seats, matching the length of the east side. These changes, along with some additional expansion on the east side, eventually brought the capacity of the wooden stands up to approximately 16,000.

In 1924 a press box was added to the west-side stands to satisfy the increasing media interest in Penn State football; however, it was a sorry affair, according to Ridge Riley. It was a shack-like structure where "the newsmen were arranged along a stationary bench with so little leg room that few could stand for the national anthem."

Almost half of Penn State's schedule of eight to nine games per season were played at home in those years. Most home games were against small Pennsylvania colleges like Geneva, Gettysburg, Grove

New Beaver Field

THE PENNSYLVANIA STATE COLLEGE

FIGURE 30 Bird's-eye view of the campus, 1910, showing relocated grandstand at the New Beaver Field location in the upper center of the view

FIGURE 31 Relocated grandstand with additional seating at New Beaver Field, ca. 1920

FIGURE 32 Grandstand plus additional seating on east and west sides, early 1920s (capacity nearly 6,000)

FIGURE 33 Construction of new wooden grandstand on the west side. The athletic grounds also included tennis courts, practice fields, and the baseball grandstand, as shown in the 1920s

FIGURE 34 With temporary seating in the end zones, 20,000 were present for a game against Syracuse University, October 25, 1924. The roof of the press box, added at the beginning of the season at the rear of the west stands, is just visible on the right margin of the photo. Varsity Hall, the athletic dormitory, stands directly across Curtin Road from the south end of New Beaver Field

City, and Carnegie Tech and drew between 2,500 and 6,000 fans. As Ridge Riley notes, the football program generally "could not persuade 'name' teams to visit isolated State College, nor could it afford to pay the required guarantees" to top-tier visiting teams.

But, over that decade, perhaps a dozen home games drew up to 20,000 spectators against opponents such as Navy (1923), Syracuse (1924), and Notre Dame (1925). To accommodate the larger crowds, especially at some of the homecoming games that began in 1920, temporary stands were erected to extend the wooden stands around the end zones, forming a complete bowl, albeit one with fewer than ten rows in the two horseshoe ends.

Penn State and Football in the Teens and Twenties

The 1908–9 move took place just as Penn State was welcoming its new leader, Edwin Erle Sparks. When it came to athletics, Sparks's position was that they "are

a necessary and important part of every well-governed American college." He believed athletics improved both the health and character of students and were vital to their enjoyment of the expanding extracurriculum and thus for success in academics. Just as he promoted student government and fraternity life, Sparks was also a football fan. While he was at the University of Chicago, coach Amos Alonzo Stagg's teams won Big Ten football championships in 1899, 1905, 1907, and 1908. At Penn State, Sparks sometimes traveled to away games with the team, and when the team played at home he often sat on the bench with the players and coaches.

His obvious affection for sports was just one of the things that cemented Sparks's popularity with the students. He was certainly a "modern" president compared to Atherton's genteel paternalism, but he was no less effective in building the success of the college. Sadly, the extraordinary pressure of operating the college through World War I took its toll on his health, and he resigned the presidency in 1920.

American participation in the war, for its relatively brief duration (twenty months, from April 6, 1917, to November 11, 1918), had a significant impact on Penn State. Beside its effect on Dr. Sparks's health, it disrupted both the work of the college and the academic progress of the students. But most sadly of all, it cost the lives of seventy-four Penn Staters. One in particular, Lieutenant Levi Lamb, was killed leading his troops in France. Lamb was a three-sport letterman who was a tackle on the undefeated 1912 football team, as well as an exceptional wrestler and 100-yard dash man on the track team. He became a symbol of all those who served and sacrificed in the war, and the fund created in 1952 that provides grants-in-aid for student-athletes and program support for all varsity teams is named in his honor.

Following Dr. Sparks's resignation from the presidency in June 1920, another dynamic leader was recruited for Penn State's top job. John Martin Thomas, whose term began in April 1921, was no less an enthusiastic promoter of the college. In fact, the trustees chose him because of his experience in fund-raising as president of Vermont's Middlebury College. There, Thomas had succeeded in gathering donations from private sources to provide much-needed buildings for "student welfare." That was certainly also the need at Penn State, but Thomas had a more ambitious plan in mind. He wanted to create a 10,000-student Pennsylvania State University, with graduate and professional schools that would be the capstone of public education in Pennsylvania.

President Thomas's $2 million Emergency Building Fund campaign, launched in 1922, became Penn State's first-ever comprehensive fund drive. Although the campaign fell short of its goal, it helped make possible the construction of both Varsity Hall in 1923 (later renamed Irvin Hall, the replacement

dormitory for the old Track House) and Recreation Hall in 1929, among other buildings.

The popularity of Penn State sports also helped promote awareness of the college. In the early stages of the Emergency Building Fund drive, Thomas wrote an essay entitled "Why I Believe in Football" for the main campaign publication. Here he showed that he was not only a proponent of college athletics; he also recognized the potency of Penn State football for drawing public attention to the college.

Unfortunately, Thomas was surprised by a contrary governor who rejected, almost in total, his grand vision for Penn State and cut its appropriations to help pay off the state's $23 million deficit. Governor Gifford Pinchot proved to be too much for the Penn State president and Thomas suddenly resigned, effective September 1, 1925, to become president of Rutgers.

The years 1909 to 1927, roughly the span of the Sparks and Thomas administrations, were the era of our first full-time coaches. William M. "Big Bill" Hollenback, his brother Jack, and Dick Harlow served in this capacity from 1909 until the shortened

FIGURE 35 President Sparks (*third from right*), on the bench with the football team. College physician Dr. Joseph Ritenour is third from left

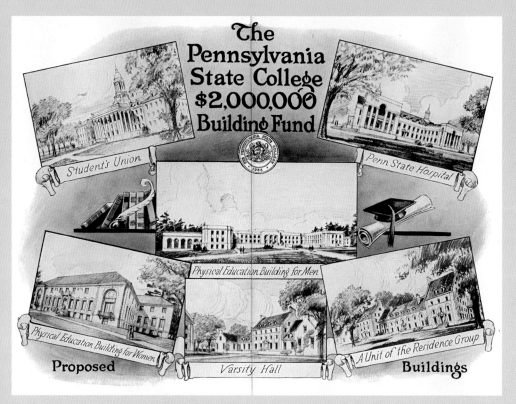

FIGURE 36 President John Martin Thomas

FIGURE 37 The Thomas Emergency Building Fund proposal included both men's and women's physical education buildings and a varsity athletes' dormitory

FIGURE 38 Hugo Bezdek coaches the team in practice in the early 1920s; note the baseball field grandstand in the background, which was adjacent to New Beaver Field at the corner of Park Avenue and North Allen Street

1918 season, when Hugo Bezdek, the manager of the Pittsburgh Pirates baseball team, was hired to take over the head coaching duties. Bezdek was no stranger to football. He had played under Stagg at the University of Chicago and coached at Oregon and Arkansas. Always a hard-driving proponent of conditioning, "Bez" was also made an assistant professor and placed in charge of physical education classes.

Overall, the Hollenbacks, Harlow, and Bezdek won about two-thirds of their games in this period. Bill Hollenback had undefeated teams in 1911 and 1912 (a 16–0–1 record overall). Among the great players of those years were All-Americans Dexter Very (second-team end in both 1911 and 1912) and E. E. "Shorty" Miller (third-team quarterback in 1913), both later College Football Hall of Fame inductees. Harlow's 1915 and 1916 teams recorded 7–2 and 8–2 seasons and, before he entered military service in 1918, his success had earned him a three-year contract extension.

Despite these achievements, losses to Pitt plagued Hollenback, Harlow, and Bezdek. As early as 1909, the Thanksgiving Day contest with Pitt had become *the* rivalry game for passionate alumni. Bill Hollenback lost his last two Pitt games, Harlow all three, and Bezdek his first in the shortened 1918 season. The six consecutive losses were assuaged with a 20–0 victory in 1919, but it would be Bezdek's last victory over Pitt in his remaining ten seasons. While his teams recorded nine winning seasons between 1918 and 1929, these Pitt losses would lead to his eventual dismissal from head coaching duties, an effort spearheaded by alumni in the Pittsburgh area.

Bezdek's 1919 team achieved its victory over Pitt in front of 40,000 fans at Forbes Field. During the teens, Penn State regularly drew between 10,000 and 30,000 fans for away games against Penn, Harvard, and Pitt.

By contrast, home-game attendance was sparse before the 1920s.

De-emphasizing Football in the Hetzel Era

With the shock of John Martin Thomas's resignation (effective September 1925), after his bold promises and the extraordinary efforts to bring them to fruition, the trustees took their time in selecting a new chief executive. Ralph Dorn Hetzel, who took office on January 1, 1927, had his bachelor's and law degrees from the University of Wisconsin, and he had spent twelve years at Oregon State as a professor and director of extension.

Hetzel came to Penn State from the presidency of the University of New Hampshire, where he had succeeded in shepherding the former New Hampshire State College to university status. This change was one of Thomas's goals for Penn State to which the trustees were still committed. Hetzel's twenty-one years in office would be among the most critical periods ever for Penn State, encompassing the Great Depression, World War II, and the unprecedented influx of veterans coming to college thanks to the new GI Bill of Rights. Despite the tremendous challenges, Hetzel's careful and patient stewardship kept Penn State from serious upheavals.

Less well known was the change that was perhaps the most dramatic in Penn State's sports history—the so-called purity movement, lasting through the 1930s and '40s. During this period, the college abandoned some of the characteristics of big-time college sports programs. Somewhat like the national experiment with prohibition, there were significant moral overtones to these changes, but over the years this "Great Experiment" became increasingly unpopular,

FIGURE 39 President Ralph Dorn Hetzel

especially with alumni and other fans who wanted Penn State to have a strong intercollegiate football program.

These policy changes took place within a context of growing national sentiment against "professionalism" in college football, in particular during the 1920s. Hetzel, along with many others, did not want to banish athletics from the campus. He believed in the value of physical education and intramurals, to provide "a sport for every man," but he recognized the problematic effects of big-time college athletics. From a student-initiated and -managed activity, it had quickly evolved into an alumni-dominated program by the turn of the century. While college administrators and faculties enjoyed the benefits of a successful sports program, Hetzel believed they had largely ignored the dilemmas football brought to higher education. He wrote:

Athletic contests became so spectacular that they quickly and in alarming measure attracted the attention of the public. Gate receipts grew, and gate receipts at institutions whose teams won the greatest number of games, exceeded the gate receipts at other institutions. This seems to make it desirable that there should be set up at each institution every possible facility designed to increase the chances of victory. Competition became keen for skilled coaches. Salaries for these men mounted in direct relation to their success in winning games. Great stadiums were built, involving millions of dollars. The time and energy of student players were so completely absorbed that they remained students in name only. Pressure was brought to bear upon institutions to offer subsidies to promising young

athletes. Serious abuses developed from which students suffered and which compromised the character and reputation of American Colleges and Universities. The spirit of good sportsmanship departed. The futility and viciousness of the whole situation began to be apparent to all thinking persons.

In the first decade of the twentieth century, Princeton University president Woodrow Wilson had declared that "the sideshows have swallowed up the circus and we in the main tent do not know what is going on." By the 1920s it more and more appeared that football was overwhelming the very foundations of academia. This led the Carnegie Foundation for the Advancement of Teaching to begin a study of the role of sports in higher education. The report, titled *American College Athletics* and issued in October 1929, was the product of three years of study and visits to over one hundred schools, including Penn State. While the report noted that colleges had made some changes, their reforms didn't go far enough to curb abuses.

Alumni Association leaders, however, were ready to make Penn State a model for reform. With Executive Director E. N. "Mike" Sullivan and Association President James Milholland providing the impetus, an Alumni Association committee chaired by John Beaver White set out to examine opportunities for reform even before Hetzel took office. Penn State had already dropped the number of athletic scholarships awarded from seventy-five to fifty. The "Beaver White Committee," however, recommended their complete elimination and several more changes. In step with the growing interest in sports for everyone was the creation of a school of physical education to both teach and oversee intramural programs; its director could not be a varsity coach. Finally, the committee recommended replacing the formerly alumni-dominated Athletic Advisory Committee with a Board of Athletic Control that included members from the Board of Trustees, the faculty, and the student body as well as alumni.

The Beaver White report appeared in March 1927, two years before the Carnegie Foundation Report and just two months after Hetzel took office. Its recommendations were approved by a majority of those alumni and students voting in their respective referendums, although opponents argued that the relatively small number of voters meant that the conclusions drawn from them were not likely representative of the larger groups. The Board of Athletic Control was implemented, and it promptly voted for the elimination of scholarships to begin in the fall of 1928. Hetzel took his time but, as the Carnegie Report was reaching its completion, he agreed to move Hugo Bezdek out of the head football coaching job by making him director of the School of Physical Education, effective July 1, 1930 (thus placating the Pittsburgh alumni and trustees who wanted Bezdek fired for the continuous losses to Pitt). Taking these moves into account, a second trustees' committee finalized the changes in January 1930, approving the last steps to both end scholarships and move athletics completely under academic and administrative control by incorporating it into the renamed School of Physical Education and Athletics. Bezdek's last coaching responsibility, that of head coach of the baseball team (which he had handled concurrently with football), ended with the spring 1930 season.

The reforms didn't just eliminate scholarships for athletes. Other practices, also common to most schools, were dropped. These included scouting opponents' games and maintaining a varsity athletes' dormitory that isolated athletes from "regular" students. Even providing tutoring sessions, a "training table" of special meals, and part-time jobs on campus to provide a few extra dollars of spending money for athletes were all considered detrimental to the amateur ideal and smacked of professionalism.

The elimination of these practices by the college administration placed Penn State in the forefront of reform. Student proponents asserted that it was "a courageous and fearless step in the right direction" and believed "the coming years will see more and more colleges and universities adopting this ideal plan." However, very few of Penn State's peer institutions followed suit. New head coach Bob Higgins, who had been one of Bezdek's All-American players, took over a football team in 1930 whose prospects were quickly changing. In Bezdek's last years, the team had experienced both ups and downs. Without college-sponsored scholarships, Higgins would manage only two winning seasons in his first ten years. From a schedule where Penn State routinely played nine or ten games, the annual number of games dropped to seven or eight, generally with only three or four home games. While Penn State was usually playing the same small Pennsylvania opponents, a schedule with fewer games meant only two of eight opponents came from out-of-state in the mid-thirties.

Average home attendance, which had reached 8,000 to 9,000 per game in 1928 and 1929, quickly dropped to a low point of 3,400 in 1934. By contrast, around 1930 most of Penn State's big-time opponents (Penn, Pitt, Syracuse, and Notre Dame) played in stadiums ranging from 23,000 to 63,000 seats. Of course, the Depression had an effect on all of these measures. The student population, for example, dipped 13 percent, from 8,600 in 1931 to 7,500 in 1933, but federal aid programs quickly helped enrollments recover and by 1936 they were back to 8,500. Fans had less disposable income to travel to home games, but

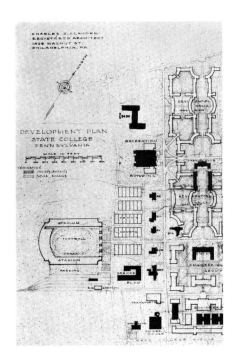 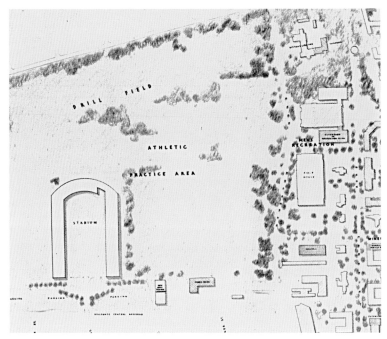

alumni in the Philadelphia area kept attendance at the annual Penn game close to 40,000 in those years.

In adopting its reform program, Penn State gave up the player benefits and competitive devices that made high-level intercollegiate athletic success possible, but it did not abandon football or ignore facilities. The administration realized that it was time to replace the wooden grandstands, which required ever more maintenance, and improve the spectators' comfort at football games and the other events held at New Beaver Field.

In the aftermath of the stock market crash, Penn State released a major development plan in 1930 to chart the future path for its building program. It was based on balancing the growth of the previous decade with meeting the economic uncertainties of the future. The work was that of Philadelphia architect Charles Z. Klauder, who had been in charge of campus planning

since 1913. The new plan encompassed completing the building groups initiated under Presidents Sparks and Thomas, removing old or temporary structures, and creating new groupings of buildings, each focusing on a particular school or activity.

Perhaps the most surprising idea was to relocate the football field and track to the south and west, next to today's bus station on North Atherton Street, then the Bellefonte Central Railroad freight terminal. Instead of the open bleachers and grandstands that had always hosted Penn State home games, there would now be a U-shaped permanent stadium, near Rec Hall and other athletic venues, that would also be close to town as well as train and bus transportation. It was the first time that a permanent structure had ever been proposed at a school that had never been able to afford more than the minimum accommodations. It was perhaps representative of the complex and

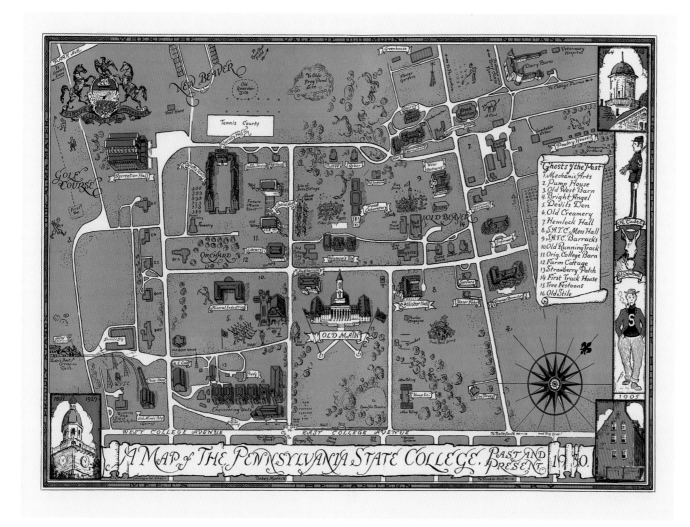

contradictory motivations of an administration then dedicated to a smaller-scale football program and an alumni and fan base that wanted to see one that was competitive with the top schools of its region, if not the country. This concept, however, remained part of campus master planning at least through 1946, with a different orientation of the stadium.

As with many other future plans, however, this change would not happen. The new stadium went unfunded. Instead, a pattern of transformation and growth began at New Beaver Field that would continue into the long-term future. The college began to replace the wooden bleachers, which had hosted athletic crowds since the 1890s, with steel grandstands. It was a change that has shaped Penn State's football venue ever since, and these first steel grandstands are still a part of today's Beaver Stadium.

FIGURE 41 Historical map of campus, drawn by art professor Andrew W. Case in 1930 for the 75th anniversary of the college issue of the *Penn State Alumni News*. It shows the location of "Old Beaver" with its nearby "Old Track House" and "New Beaver" with "Varsity Hall." For New Beaver Field, the slightly widened boundary midway along the rear of the west stands reflects the presence of the press box

NEW BEAVER FIELD
THE STEEL ERA

New Beaver Field's steel era began in 1934, when the planned replacement of all wooden stands and bleachers with steel grandstands commenced. The Depression necessitated a gradual approach, however, and the conversion to steel stands was completed in four phases over the years from 1934 to 1939. This brought the field's seating capacity to 14,700.

By contrast, around 1930, Penn State opponents Notre Dame, Penn, Pitt, and Syracuse played in stadiums ranging from 23,000 to 63,000 seats. Six of Penn State's future Big Ten opponents (Illinois, Michigan, Minnesota, Ohio State, Purdue, and Wisconsin) had stadiums ranging in capacity from 26,000 to 86,000 attendees. A further expansion of New Beaver Field in 1948–49 almost doubled steel grandstand seating to 27,720, but this was considered the maximum capacity for the site.

Plans for the Athletic Plant Take Shape, 1920s–1930s

The campus envisioned in the long-range plan of 1907 had placed athletic activities in the northwest quadrant, and college architect Charles Klauder's plans for a gymnasium and a relocated athletes' dormitory were part of President Thomas's 1921 Emergency Building Fund campaign. As noted earlier, when Hugo Bezdek was hired in 1918, he was appointed director of physical education in addition to coaching football and baseball. He also shared the duties of athletic director with the graduate manager of athletics, the alumnus who controlled athletic finances. In these roles Bezdek would play a critical part in the expansion of campus athletic and recreational areas.

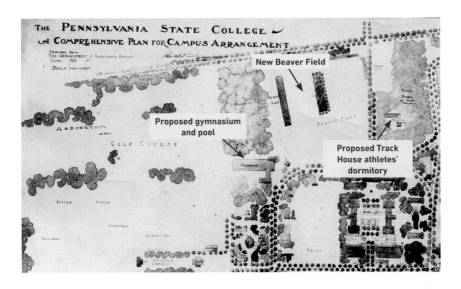

Within the image:
THE PENNSYLVANIA STATE COLLEGE
A COMPREHENSIVE PLAN for CAMPUS ARRANGEMENT

New Beaver Field

Proposed gymnasium and pool

Proposed Track House athletes' dormitory

ARBORETUM AND
GOLF COURSE

The 1921 Emergency Building Fund campaign put new dormitories and physical education facilities for both men and women on the physical plant priority list. Varsity Hall, which replaced the old Track House, was relocated from Hort Woods to become the initial structure in a new West Halls quadrangle, where it is now known as Irvin Hall. Funds for this building, at the insistence of Hugo Bezdek, came from the football team's $25,000 share of the 1923 Rose Bowl receipts, along with additional contributions from Emergency Fund donors. Further evidence of the commitment to the development of the athletic plant at the northwest corner of the campus came when the wooden grandstands were systematically replaced by steel.

FIGURE 42 Plan for landscaping the campus shows proposed Track House athletes' dormitory in Hort Woods, 1921. The gymnasium and pool are in the approximate Rec Hall location, with outdoor sports fields to the west along the golf course

The developing plans for these concepts were seen as early as a 1920 planning study for landscaping the campus. The detail of this plan (shown in fig. 42) placed the gymnasium (with an attached pool) in its eventual location. The Track House was envisioned in the middle of Hort Woods. Football and baseball grandstands remained in place, and tennis courts on the southern edge are indicated. Also detailed in this design were a golf course and arboretum to the west, just below Park Avenue, and farther south extensive spaces for soccer and baseball fields, volleyball and basketball areas, more tennis courts, and an "outdoor gymnasium" for track, along with structures identified as shelters, a fieldhouse, and a clubhouse (perhaps for the golf course). Much of this space was eventually used, however, to enlarge the golf course to eighteen holes. The realization of these additions to recreational areas can be credited to Hugo Bezdek, who in his role as a professor in charge of physical education classes was committed to broadening both sports and intramural programs to benefit the student population.

The Transition to Steel, 1934–1939

Beaver Field's new steel grandstands were a patented product referred to as a "Lambert Grandstand." Byron J. Lambert, who held this patent for "grandstand construction," was at one time professor and head of civil engineering at the University of Iowa. This innovation used prefabricated sections of steel decking that were bent to form the stepped rows of the grandstand. The sections of decking were then bolted together to form whatever configuration of the number of rows and the length of the stands desired. These decks were supported underneath by steel columns with cross-bracing. The work was then completed with the installation of steel supports to which wood bench seating was attached.

The 1934 phase of steel construction took place on the west side, replacing the wooden stands there with twenty rows of steel stands, 180 feet in length, which provided seating for approximately 2,160 fans (see the color-coded summary of the construction stages

in fig. 54, item 1). The work began on September 4 and was quickly completed in time for the first home game of the season against Lebanon Valley College on October 6, 1934. This initial section was installed slightly off-center from the fifty-yard line to accommodate the future expansion plans that would create a symmetrical layout from the midfield marker. The remaining seats used in the 1934 season were wooden grandstands on the east side, providing a total seating capacity of about 5,500.

The second addition was erected between the 1935 and 1936 seasons. It was an identical 180-foot section of steel grandstand and was erected on the east side to match that on the west side (fig. 54, item 2). This increased the total seating capacity of the steel grandstands to approximately 4,320, while some of the wooden structure remained on the east side to augment the more permanent steel construction.

The third increment began in 1937 when, as a first step, the steel stands on the west side were lengthened to 306 feet and expanded to forty rows (fig. 54, first part of item 3). They were centered on the fifty-yard line and now extended from goal line to goal line

with a few feet to spare. Figure 43 (taken from the balcony of today's Irvin Hall) shows the ongoing construction of the expanding west-side grandstand and the existing east-side grandstand, and figure 44 shows the completed configuration of the 1937 west-side construction. This figure also shows the press box, first added in 1924, which had been retained with only minor improvements for the press, although it was extended to provide covered seating for some college officials and their guests.

Also in 1937, after the completion of the west-side expansion, the east side was lengthened (see the second part of item 3 in fig. 54). These changes brought total seating capacity for the steel grandstands to over 11,000. Figure 45 also shows the chain-link fence that separated the running track from the pedestrian walkway and the concrete ramps that provided access to the seating areas. This arrangement would be carried around the north end when New Beaver Field was expanded again in 1949, and was also used for the initial configuration of Beaver Stadium, albeit with a wider asphalt walkway between the fence and the ramps.

FIGURE 43 On the left is shown the 1937 expansion of the west-side grandstand to 306 feet long, forty rows, which nearly tripled the size of the initial steel "Lambert Grandstand" on the west side (180 feet long, twenty rows, built in 1934). On the right is shown the initial portion of the steel stands on the east side (180 feet long, with twenty rows), built in 1935–36

FIGURE 44 West stands, 1937 (306 feet long, forty rows). The Water Tower, constructed in 1936, is also shown in this 1941 photo

Figure 47 Aerial view of New Beaver Field,
1941 (seating capacity of 14,700)

Figure 45 Extended east-side grandstand
(306 feet long, twenty rows), 1937

Figure 46 East stands, 1939 (306 feet long,
40 rows), as shown in this 1941 photo

Some temporary wooden grandstands that augmented the initial steel structure on the east side were demolished when the Board of Trustees appropriated funds in 1939 for constructing the fourth and final expansion of the steel grandstands and repainting the entire structure. This work doubled the height of the east stands, making it forty rows to match the west stands (see item 4 of fig. 54).

Now the grandstands on both sides consisted of forty rows of seats extending for the full length of the football field, from goal line to goal line. This would remain the configuration of New Beaver Field until 1949. At this point, the seating capacity was approximately 14,700, a modest reduction from 16,000, which was the largest seating capacity of the last set of wooden grandstands. Of course, temporary wooden bleachers could still be installed at both the north and south ends to provide additional seating when needed for high-demand events like the annual homecoming game, as had been done during the wood era. The "electric" scoreboard was located at the north end of the field adjacent to Park Avenue. It was installed there in 1937 to replace the earlier manual scoreboard and was a gift of the class of 1926. It would be moved to the south end in 1949, to stand between the two flagpoles, a gift of the class of 1944, when the horseshoe was added at the north end.

All of the original design drawings for the changeover from wood to steel identify the new steel structure as using the patented Lambert Grandstand, with the work done by the Pittsburgh–Des Moines Steel Company. The use of this type of grandstand construction made it possible to later relocate these stands. As will be described more fully in a later chapter, all of the steel grandstands present in 1949 were moved and incorporated into Beaver Stadium in 1959–60.

New Beaver Field, like its predecessor, was designed to accommodate both football and track (see fig. 47). The circumferential running track along with its west-side corner extensions provided straightaway sections for short-distance running and hurdling events. The spacing between the edge of the playing field and the stands was slightly greater on the west side to accommodate the wider running track with lanes for runners in the dashes and hurdling events.

New Beaver Field (and later Beaver Stadium as first configured) was thus a "multipurpose" field, with football being played inside the track. Typically stadiums designed exclusively for track and field used the turfed area inside the track for the jumping and throwing events. These included tracks and sand pits for long and triple jumps and for pole vaulting. A high-jump area would occupy one end of the infield. Throwing events would then typically begin at the other end from different starting positions. The shot put, javelin, hammer, and discus throws all used the long dimension of the infield parallel to the straights.

Since New Beaver Field's football playing surface required some degree of protection against overuse (there were separate football practice fields adjacent to the main field), the high-jump area and spaces for throwing events were situated outside the south end of the field, between the football end zone and the tennis courts along Curtin Road. While track and field competition at Penn State preceded even Old Beaver Field's construction, New Beaver Field offered a more modern track and field facility. By the early 1920s Penn State's track team was becoming a national power with record-setting performances in the middle- and long-distance runs and hurdles and in the hammer throw. In 1922 Penn State placed second in the national track and field championships behind the University of California, and a number of Penn State athletes placed highly in a variety of track and field events in the 1920, 1924, and 1928 Olympic Games.

When the west stands were expanded to forty rows in 1937, plans were also drawn up to enclose the underside of the stands with an exterior brick wall to create team rooms, handball courts, and other facilities. While this work was not done at that time, water lines and restroom facilities were added to both sides over the years. In 1937 the Board of Trustees appropriated funds to locate team rooms in the second and third floors of the new water tower, which was constructed directly behind the west stands in 1936.

De-emphasis and Stadium Changes in the 1930s

The new steel grandstands initially had a smaller capacity than the wooden bleachers that they replaced. Even after the seating expanded to 14,700 in 1939, it was still less than the 16,000 of the wood era, not to mention the occasional games where, with temporary bleachers, capacity could reach more than 20,000. Did the decision to de-emphasize football, advocated by President Hetzel with alumni, student, and trustee concurrence, play a role in this decreasing of the size of the grandstands? The elimination of scholarships and recruiting as well as other "professional" perks like training table meals, a separate athlete dormitory, and special tutoring to help players maintain eligibility certainly had an impact on the quality of players coming out for football. The scope of the football program was also diminished by having fewer games, fewer out-of-state opponents, and fewer quality opponents. Despite a less-challenging level of competition, the result was nearly a decade of losing seasons for the football team. And yet, the success of other Penn State athletic teams outside football seemed to remain largely unchanged. The loss of scholarships primarily affected football, since the other sports had fewer scholarship players.

The 1930s began with eleven sports at Penn State: football, soccer, and cross-country in the fall; basketball, boxing, wrestling, and track (indoor) in the winter; and baseball, lacrosse, golf, tennis, and track (outdoor) in the spring. Over the decade, swimming, fencing, gymnastics, skiing, and rifle also joined the lineup and there were freshman teams in football, basketball, baseball, cross-country, and track.

One of the great innovations of the Hetzel era was the goal of "having every student in some sport." This may have been an outcome of the discovery that one-third of America's young men were physically unfit for military service in World War I. The intramurals program drew over 3,500 male students into participation in thirteen team and individual sports. Women's sports were also popular and were administered separately under the Women's Athletic Association. There, a large number of coeds were involved in archery, baseball, basketball, fencing, field hockey, golf, hiking, riding, rifle, skating, skiing, tennis, tobogganing, track, and volleyball.

In men's sports, it was an era of legendary Penn State coaches. Besides Bob Higgins in football, there were Bill Jeffrey in soccer, John Lawther in basketball, Leo Houck in boxing, Charlie Speidel in wrestling, Nick Thiel in lacrosse, Joe Bedenk in baseball, Bob Rutherford in golf, and Nate Cartmell and then Chick Werner in track, with Werner also coaching cross-country. Track and cross-country teams were outstanding in this period, turning out first-place winners in the IC4A championships and the Penn Relays as well as several Olympians. Wrestling had undefeated seasons and top ranks in eastern intercollegiate championships. Boxing teams were routinely intercollegiate champions, and Bill Jeffrey's soccer teams had an unbeaten streak of sixty-five games beginning in 1932, with eight of his teams between 1926 and 1940 considered national champions. Large crowds for winter sports were not uncommon—boxing, wrestling, and basketball would sometimes bring attendance in Rec Hall to as many as 8,000 with temporary seating and standees to accommodate such an overflow crowd.

Most of Penn State's non-football teams continued to compete against the same set of regional institutions with some variations. Early-season games for these sports would often come against schools such as Bucknell, Dickinson, Gettysburg, and Washington & Jefferson. The "meat of the schedule" would then include teams such as Army, Cornell, Navy, Penn, Pitt, and Syracuse, with schools such as Colgate, George Washington, Georgetown, Harvard, Lafayette, Lehigh, Princeton, Temple, and West Virginia sprinkled into the schedules. Intersectional games for non-football sports were rare, but occasionally a Penn State team would meet a Midwestern opponent like Iowa, Michigan, or Wisconsin, or southern schools like North Carolina or Virginia. There were also some games that were unique to particular sports, such as annual lacrosse matches against the Onondaga Indians, and soccer and lacrosse games with the University of Toronto. Rec Hall hosted several eastern intercollegiate championships in boxing and wrestling in the thirties and the NCAA wrestling championship tournament in 1933.

Given the level of success Penn State's non-football sports continued to maintain, and the excitement varsity sports and intramural competition continued to generate among the student body, it is clear that the policies meant to "de-emphasize professionalism in football" had little impact on other sports at Penn

State. We can only conclude that football was a special case. Nationally, it remained the most popular of college sports, and although Penn State had stepped back from the practices that seemed to define "big-time college football," it found itself largely alone in embracing this noble but ultimately futile gesture.

Expansion of the Athletic Plant, 1930s–1940s

Solicitation of donors for a long-needed new gymnasium had begun in 1918, but it wasn't until after the 1925 completion of the Emergency Building Fund campaign that construction began in 1927, taking about eighteen months. The building was dedicated when it hosted the tenth national Intercollegiate Boxing Championships on March 22–23, 1929.

As Penn State's athletic contests became more popular with alumni and fans from out of the area, it became obvious that more overnight accommodations were needed. In addition, extension and research activities were generating an increased demand for hosting conferences and meetings that could not be provided for off-campus. So in May 1930 the college broke ground for a new hotel, located in the woods next to New Beaver Field and Rec Hall. By 1931 it opened with rooms for $3.50 a night and meals offered in a traditional, informally furnished country inn. The Nittany Lion Inn was originally under the management of the Treadway Corporation. It became a self-supporting arm of the college and a training ground for hotel management majors in 1948.

Another landmark of the athletic area of campus was a new campus water tower, constructed in 1936 behind the west grandstands of New Beaver Field. Ordinarily such a utilitarian structure would stand almost unnoticed. But enclosing the space below

FIGURE 48 Steel work and rising brick walls
for Recreation Hall, 1928 (TOP);
football equipment room in Recreation Hall,
ca. 1930 (BOTTOM)

FIGURE 49 Nittany Lion Inn construction,
1931

FIGURE 50 Water Tower, 1938; note New
Beaver Field grandstands and press box
behind

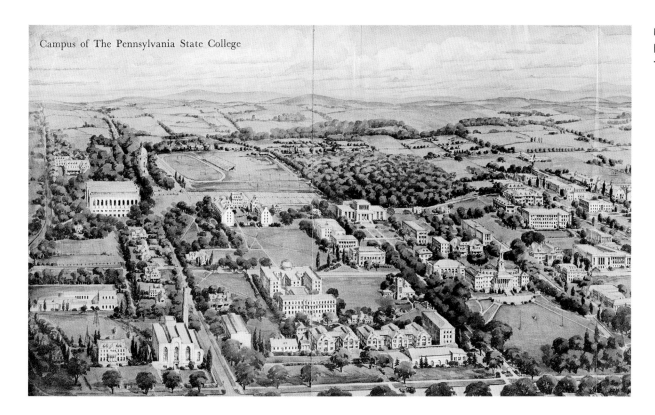

Campus of The Pennsylvania State College

the water tank in brick was an unusual step taken by campus architect Charles Klauder that complemented the architectural style of Recreation Hall and the West Halls quadrangle. It was designed to provide locker rooms, medical and equipment spaces, and showers and bathrooms on both its second and third floors. Thus it formed an integral part of the athletic plant well before locker rooms were built under the grandstands after World War II.

While it didn't directly benefit athletics, the Nittany Lion statue, created in 1942, was the most celebrated feature of the area. The gift of the class of 1940 is today one of the campus's most treasured landmarks. Sculpted by Heinz Warneke (see fig. 68, chapter 5), a renowned artist famed for his representations of animals, it was placed in a small wooded

grove between Rec Hall and New Beaver Field, next to the field's ticket booths. It faced a parking lot that, in the view of President Hetzel, could serve as a gathering place for student rallies and bonfires, moving celebrations away from the downtown corner of College and Allen streets. There impromptu bonfires and other student "exuberance" in the past had damaged both street bricks and parking meters, not to mention friendly relations with the town officials.

Horseshoe and Extensions, 1948–1949

The culmination in the expansion of the athletic plant at the northwest corner of campus was announced in August 1948. A pen-and-ink conceptual sketch (fig. 52) by Walter Trainer, an assistant professor of landscape construction who also directed landscape planning and maintenance for the college, appeared in the *Centre Daily Times* showing how the expansion of New Beaver Field would increase the "permanent seating capacity from 14,700 to approximately 28,000." The college could anticipate rising attendance at football games; Coach Higgins's teams had ten consecutive winning seasons from 1939 to 1948. Since the end of the war, average home game attendance had continued to rise, and three of four home games in 1948 exceeded the grandstand's capacity, with 24,579 crowding the 14,700-seat facility at the homecoming game versus Michigan State.

Approval for this expansion occurred through a sequence of separate actions by the Board of Trustees. Initially, in January 1948, the Board of Trustees' Building and Grounds Committee recommended to the full board that the seating be expanded with "additional steel stands on the east and west sides of the field." This involved a seventy-two-foot extension of forty rows of seating on the south end of the stands on both sides of the field. This would add a total of 3,460 seats and push the stands beyond the south end

zone. It was the intent of the board to proceed with this phase of the expansion prior to the 1948 season; however, construction was prevented because of a national shortage in the supply of steel.

Then, in December 1948, the previously approved seventy-two-foot expansion on the east and west sides was augmented to include "construction of additional stands with a capacity of 9,400 at the north end of the field." These improvements, some of which were already "under way" according to the meeting minutes, were "authorized" in January 1949 for completion by September 30, 1949.

The north-end expansion transformed New Beaver Field into a horseshoe form and actually provided 9,560 seats by attaching the thirty-row curved section directly to the north ends of the existing stands. The proximity to Park Avenue prevented construction from exceeding thirty rows at the north end of the field. A unique feature of New Beaver Field after the curved section was added was the necessity for an opening in the stands at the northwestern corner to accommodate the continued use of the elongated straight extension of the running track. Once again, the steel construction for this addition was done by Pittsburgh–Des Moines

FIGURE 52 Announcement of the 1949 expansion of New Beaver Field, August 20, 1948. Sketch by Walter Trainer

FIGURE 53 New Beaver Field, early 1950s (seating capacity 27,720)

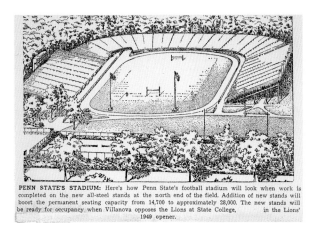

PENN STATE'S STADIUM: Here's how Penn State's football stadium will look when work is completed on the new all-steel stands at the north end of the field. Addition of new stands will boost the permanent seating capacity from 14,700 to approximately 28,000. The new stands will be ready for occupancy when Villanova opposes the Lions at State College, in the Lions' 1949 opener.

Steel Company and the new structure was "manufactured and/or licensed for use under one or more of patents" held by Byron J. Lambert.

With the seventy-two-foot extensions to the south end of both the east and west grandstands, and the addition of the horseshoe at the north end, the total number of seats added was 13,020, which increased the seating capacity of New Beaver Field to 27,720 (shown in fig. 53 and item 5 in fig. 54). The January 1949 board action also included authorization for the expansion of the press box, which was reconfigured to have four levels, but with a narrower base. Three levels were evident from inside the stadium, but the fourth level, which housed a kitchen and restrooms, was hidden from view behind the west stands. In addition, a visiting team locker room was constructed under the west stands.

A color-coded diagram of New Beaver Field (fig. 54) shows the incremental development of the steel grandstand to its final permanent seating capacity of 27,720. This could also be augmented by temporary wooden bleachers at the open south end that would increase the capacity by 2,295. The first game for the enlarged venue, against Villanova on September 24, 1949, had an attendance of just 22,080. It only took until homecoming 1951 to surpass the new capacity figure when 30,321 saw Penn State fall to Michigan State by a score of 32–21. Clearly, athletics administrators believed New Beaver Field's expanded capacity was sufficient to meet the need in 1949, but those attitudes would soon change.

Football Revives from the 1940s into the 1950s

Along with the challenges of weathering the Great Depression, Penn State had taken its dramatic step

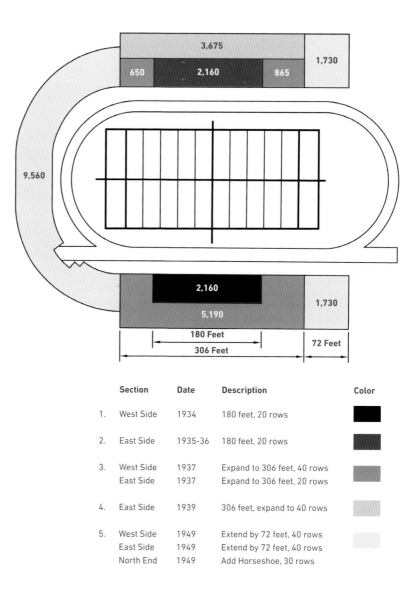

	Section	Date	Description	Color
1.	West Side	1934	180 feet, 20 rows	
2.	East Side	1935-36	180 feet, 20 rows	
3.	West Side	1937	Expand to 306 feet, 40 rows	
	East Side	1937	Expand to 306 feet, 20 rows	
4.	East Side	1939	306 feet, expand to 40 rows	
5.	West Side	1949	Extend by 72 feet, 40 rows	
	East Side	1949	Extend by 72 feet, 40 rows	
	North End	1949	Add Horseshoe, 30 rows	

FIGURE 54 New Beaver Field expansion in steel era (numbers represent number of seats). Drawing design by Harry H. West

toward reforming intercollegiate athletics by elim-
inating athletic scholarships, recruiting, and other
benefits to athletes. However, few of its peer institutions chose to follow this lead and, at least in football,
the college suffered the consequences; Higgins's
teams went nine years, from 1930 to 1938, without a
winning season, severely testing student, alumni, and
fan enthusiasm.

After two embarrassing losses to Waynesburg
College in 1931 and 1932, Pittsburgh alum Ben C.
"Casey" Jones and a group of fellow grads combed
high schools in steel and coal towns for talented
young men. They had little to offer, but could promise
summer jobs, perhaps free room and board in a fraternity house, and a part-time job during the year.

By 1936 individuals on the Board of Trustees were
encouraging alumni to "give aid to worthy athletes" and
help find them part-time jobs, while still maintaining
the no-scholarships policy. They slowly made progress,
and football in 1939, '40, and '41 began to return to winning ways, compiling an 18–4–3 record in those years.

Athletics was thrown into turmoil by World War II.
Many upperclassmen left school for the armed forces.
Higgins continued to fashion winning seasons with
what young talent he had, although in the words of one
commentator, "college football was only a shadow of its
former self." Starting in 1945 an influx of older, mature
athletes, returning to college under the GI Bill, seasoned
several years of rosters. At the same time, freshman
football players were "farmed out" to California State
College due to lack of housing on the Penn State
campus. There Coach Earl Bruce developed a growing
number of talented young athletes for Penn State.

The undefeated 1947 season, which culminated in
an appearance in the 1948 Cotton Bowl, Penn State's
first bowl game in almost twenty-five years, was symbolic of the drawing back from the reformist policies
of the early Hetzel years. Alumni continued to recruit
and offer financial support to star players. Finally, in
1949, the Board of Trustees lifted the ban on financial
aid and instituted 100 grants-in-aid for athletics, of
which football received 48.

The following year, the new Penn State administration of President Milton S. Eisenhower increased total athletic scholarships to 150. But under this new leadership, there was an insistence that Penn State football, and athletics in general, would recruit outstanding players who were qualified academically to be regular students and who could, and would, graduate on time.

With the arrival of the 1950s and the Rip Engle era, Penn State football began to improve its standing in the football world while maintaining a balance between athletics and academics. This new era saw increasing athletic success and substantial increases in the numbers of fans who wanted to enjoy Penn State football on campus. The expanded New Beaver Field soon proved to be inadequate to meet the ticket demand for all those who wanted to come see a game. It would not be long before a rapidly growing Penn State had to, once again, consider how and where to play its home football games.

5

FOOTBALL GAMES AT NEW BEAVER FIELD

Student life in the teens and twenties reached the high point of "collegiate culture," where the image of attending college became a lifestyle portrayed in movies, music, magazines, and other forms of popular culture. Probably nothing matched the football game for bringing all the parts of this exciting experience together. With the stock market crash in October 1929, the Great Depression, and the gradual descent into World War II, changes in college life took hold but the old style and enjoyment were not completely lost. After the war, the GI Bill of Rights made it possible for more veterans to seek degrees, and these older students came to college with the feeling of lives postponed. This new sense of purpose allowed little time for the old-style juvenile college culture. By the 1950s, customary student life, updated for a new era, had returned, but soon Korean War veterans were

returning, thanks to extended GI Bill benefits, and they too had no time for "college nonsense!" Through all these changes, football was a constant presence.

Going to the Game

From the 1920s to the 1950s, many changes came in the collegiate culture that had come to fascinate the public. Not the least of these elements was how students dressed. The dominant theme appeared to be the evolution of casual clothing as part of the student wardrobe, which also came to play a major role in American fashion for all age groups. Students presented a diversity of dress. As a public land-grant college, Penn State's students ran the gamut of socio-economic backgrounds, rural and urban origins, and

Crowd photos from games of these years show people more formally attired. Penn State alum and faculty member Mickey Bergstein described this era:

The earlier crowds dressed up to attend a football game on the campus. The women mostly wore suits, dresses. In addition, most of the women wore hats which were a part of their "outfits." . . . The men on the other hand, generally wore suits or a sports jacket and many of the men . . . were wearing fedoras. . . . The other feature of women in the stands, both older women and if a college girl had a date for a game, and if her date could afford $1.50 or $2.00, was the wearing of a pompom, with the team's ribbon a part of the flower arrangement.

Even in the 1960s a man could still buy a white football mum corsage for his date on the way to the game.

From the 1920s on, men's dress on campus became more casual, and by the late fifties it was rare to see a male student in a coat or sweater with a tie. While a fraternity man might wear a suit and tie to a football game, an independent man would more likely dress "respectably, but casually." Women students' clothing standards over these decades were fixed by college rules as much as by fashion conventions. In the Depression one might assume that clothing budgets would have been tighter, but in the words of Grace Holderman (class of '32), students of that era "dressed well, collegiately well that is." The typical Penn State student may have lacked the wardrobe of a Princeton man or Vassar woman, but some standards were observed (or at least aspired to).

Getting to the game was an easy walk, no more arduous than going to class. The location of New

FIGURE 56 New Beaver Field crowd, ca. 1930s

FIGURE 57 Students at the game at Beaver Stadium, early 1960s

academic majors. How they dressed depended on the occasion, whether it was for class, a laboratory session, a formal dance, casual relaxing, or going to a sporting event. But football games seemed to be different.

Beaver Field put it in close proximity to the half-dozen on-campus fraternity houses as well as the "Tri-dorms"—the only men's residence halls. These three men's dormitories in the 1930s and early '40s—Varsity (Irvin), Watts, and Jordan halls—housed about 300 men. A large majority of fraternities, however, were located in town, and before World War II most independent men lived off-campus as well, in private rooming houses scattered through town. Coeds lived in the several women's dormitories—the Women's Building, McAllister Hall, Grange Hall, and Atherton Hall, as well as in the few sorority houses, which were all on campus (former faculty residences converted to this purpose). For most games, the gentleman picked up his date at her dorm or house and walked with her up to the field for the game.

The spectators, Bergstein noted, "were mainly students and alumni, who were joined by residents of State College and the nearby communities. The University staff also was included in the usual crowds at the Penn State games." With the exception of a special occasion like homecoming or "Dad's Day" (when the fathers of students were invited to join their son or daughter at a game), people in the twenties and thirties rarely drove from very far away to a game. Paved roads were uncommon in the 1920s, and long-distance auto travel during the Depression was usually limited to necessities. Passenger rail service to town via the Bellefonte Central Railroad ended in the early 1920s, except for the rare special train, and commercial bus service began to replace it. In addition, Pennsylvania Railroad passenger service to Lemont slowly disappeared through the 1930s. The spectators for most home games before the 1950s were largely a hometown crowd.

During World War II travel restrictions resulting from gasoline rationing kept crowds relatively small,

but with the end of the war, restrictions soon eased and attendance figures at football games began to rise. More people were coming to games from out of town, especially alumni. By the late 1940s seats at games were in high demand, and complimentary tickets for spouses of faculty, staff, and veterans were eliminated, as were those for visiting high school football teams. With the near-doubling of the seating capacity in 1949, ticket demand was largely satisfied for a time, but with the team's success under Coach Rip Engle, demand continued to rise through the 1950s and more people were driving increasing distances to games.

There was no parking on the adjacent practice fields, the baseball field, or the tennis courts, but the comparatively small parking lots in front of Rec Hall and by the Nittany Lion Inn were filled. In addition, there are numerous accounts of youngsters from the 1950s who organized parking in the driveways and

FIGURE 58 "Tri-Dorms"—Varsity (Irvin), Watts, and Jordan halls

FIGURE 61 Students gather at the Corner Room

FIGURE 59 Celebratory dinner for fraternity and alumni, ca. 1920

FIGURE 60 Pregame pep rally with the Nittany Lion, ca. 1950s

front lawns of their homes in College Heights to earn extra money on an autumn Saturday. Yet by comparison to the 1960s and beyond, "there were no game parking lots to hold thousands of cars and pre-game and post-game partying," as Mickey Bergstein put it.

Homecoming was a special day for all who attended. In the 1920s Friday nights would feature an "athletic mass meeting"—we might call it a pep rally—in Schwab Auditorium featuring the football team, alumni cheerleaders, and the Blue Band playing the college songs. Saturdays and Sundays featured open houses, the annual Horticulture Show, an all-college cider party in the Armory, and Saturday night reunion dinners. By the thirties a golf tournament was added for Saturday morning. On Sundays the Penn State Christian Association sponsored breakfast in the Old Main basement sandwich shop before chapel services in Schwab Auditorium.

As more alumni returned for homecoming, more events had to be planned to keep them occupied. In the post–World War II period, there was an expanded alumni luncheon in Rec Hall on Saturday, theater presentations by both Penn State Players and the Thespians, and an alumni dance in Rec Hall Saturday night.

"Pennsylvania Day" in the teens and twenties was a special day when the governor, legislators, and other government officials visited the college and attended a football game. But the weekend students most looked forward to was the "Fall Houseparty." John P. Ritenour '39, the son of campus physician Joseph P. Ritenour and a Kappa Sigma member, recalled, "They were always gala affairs, with everyone dressed in their best. . . . There was always music and dancing, and formal dinners at the house—these were romantic times! House Parties were scheduled when there would be football games, so all the guys took their dates to the games on Saturdays, then returned to the house for dinner and dancing."

On most home-game Saturdays fraternity members returned to their houses for dinner after the game, while independents could eat at the Old Main Sandwich Shop, at their boarding houses, or at familiar downtown spots like the Corner Room. Alumni and visitors might go to the Nittany Lion Inn for dinner (after it was built in 1931), or to a downtown restaurant, or even to a banquet at the Centre Hills Country Club.

By contrast, attending an away game was a more complex endeavor for a Penn State student. In earlier times the president and faculty limited the number of away games students could travel to or restricted them to Pennsylvania locations. In the 1930s and '40s, the bulk of road games were with Pennsylvania teams or with schools in adjacent states, especially New York. Penn State played at Cornell, Syracuse, Army, Colgate, NYU, Columbia, and Fordham in those years, with one Colgate game staged in Buffalo. These were games that students and alumni could travel to, unlike the rare game at Michigan State, South Carolina, or Washington State. For example, arrangements for the 1933 Columbia game in New York City had been made by Penn State's student union; tickets and other reservations could be purchased at the offices in Old Main. A game ticket was $2.20; a round-trip bus ticket to the city was available for $7.50 or a combination bus and train ticket for $8.75. Fraternity members could stay at fraternity houses at Columbia with dancing after the game or at the Hotel Times Square for $5.50, which covered a room, meals, entertainment at Radio City, and a sightseeing trip. In 1933, the depth of the Depression, a $20 weekend was a significant luxury; calculating for the rate of inflation, the equivalent amount today would be about $360!

FIGURE 62 Wilfrid Otto Thompson

Pennsylvania games were likely more afford-able. Today we recall the annual Thanksgiving Day game against Pitt as the main rivalry game of the year. However, before the 1950s, the annual game against Penn was just as important to alums in east-ern Pennsylvania. John Ritenour recalled, "Another fond Philadelphia memory is of going to Philly for *the* football game—Penn State versus the University of Pennsylvania. It was played every year in Franklin Field—they got much bigger crowds than State College could draw—and were usually very good games. Penn State student and alumni activity always centered around the Bellevue-Stratford Hotel" in Center City; the "Grande Dame of Broad Street" was the social center for the Philadelphia establishment.

Penn State played Penn forty-seven times between 1890 and 1958, although the competitiveness of the games began to decline after Penn agreed to join the other Ivy League schools in de-emphasizing athletics after 1951. In the early years, before 1911, Penn State was 0–16–1 against Penn, being outscored 386 to 24 points; by the 1940s and '50s, Penn State dom-inated Penn, winning nine of the last ten games they played. Franklin Field was one of the largest stadiums in the country, and from 1925 through 1953 Penn State games there routinely drew 40,000 or more fans, sur-passing 71,000 in 1948.

The first away games against Pitt, then the Western University of Pennsylvania, were held at Exposition Park, the baseball field on the north side of the Allegheny where the Pittsburgh Pirates originally played. By 1909 the games were moved to Forbes Field in Oakland, adjacent to the new loca-tion of the renamed University of Pittsburgh. There Penn Staters traditionally made the Schenley Hotel on Forbes Avenue their headquarters. "Pittsburgh's class hotel of the early 20th Century" was a block

away from Forbes Field. In 1925 the new Pitt Stadium opened, less than a mile from the hotel but including a climb up "Cardiac Hill" to get there. Unlike Penn games, attendance for games at Pitt rarely exceeded 30,000 until after World War II; except for rare occa-sions, the game was always played in Pittsburgh.

On and Off the Field

In 1914 President Sparks hired Wilfrid Otto Thompson, known to most Penn Staters of the time as "Tommy," to be the college's bandmaster. Thompson had over twenty years of service directing Army bands and he was a colleague and close friend of John Philip Sousa. He had high expectations for Penn State's cadet band. He took control of the band from its student officers, and it was officially made a part of the Military Science Department.

The band had sixty members, still wearing their Civil War–style union blue uniforms. After World War I the band doubled in size and resumed performing at home football games, and also at two away games with fewer musicians. In those days the band would march around the track before the game and then remain in the stands except for halftime. The shows consisted of marching in military style and sometimes forming numbers or letters (typically a block "S").

By the early 1920s the band was one of the largest college bands in the country. An expanding budget permitted the purchase of fifty new blue uniforms for the best musicians. This "Blue Band" was the group that traveled to away games, playing for both alumni events and then at the game on Saturday. In Pittsburgh they were featured on radio station KDKA, along with the glee club and talks by President Thomas and Coach Bezdek. These broadcasts reached listeners over most of the United States and eastern Canada, and could also be heard as far away as London and Paris.

In 1929 the band, all uniformed in blue, became part of the new School of Music. After Thompson's retirement in 1939, Professor Hummel Fishburn took over. He gradually did away with the military style, making the Blue Band more of a collegiate band. They performed more intricate and innovative maneuvers for halftime shows, with a new 180-step-per-minute marching style still used today to enter the field at the beginning of a game. During World War II the band drafted military trainees temporarily stationed at the college, women students, graduate students, faculty, State College High School musicians, and community musicians to make a band, which dressed in street clothes or military uniforms and only played from the stands.

After the war, the Blue Band quickly made a comeback paralleling the football team's success. Their greatest disappointment was not being allowed to join the team at the Cotton Bowl in 1948, as the college determined it needed the $75,000 in bowl proceeds to cover a badly out-of-balance athletic budget. In 1949, the band's fiftieth-anniversary year, James W. Dunlap took over the Blue Band and brought a new era of music to football game spectators. He continued as director into the Beaver Stadium era. Dunlap instituted Band Day in 1950, with twenty-four high school bands, each directed by a Penn State alum, playing in a "PSC" formation on the field at halftime.

Also on the field, cheering became a more organized activity in this period: a squad of five or more with a head cheerleader and, still, a second-in-command known as the "song leader." Photos of games of the twenties and thirties show the cheerleaders along both sidelines at New Beaver Field. Since visiting teams usually did not bring many supporters all the way to State College, spectators in the grandstands were primarily Penn State fans, including students, alumni, and community members. The cheerleading squads were not large, so individual members were spread as far as twenty or more yards apart to cover the entire crowd, rather than being bunched together in front of the student section as they are today.

The Lion mascot had made only sporadic appearances in the 1920s. It was not until 1939 that School of

FIGURE 63 Band Day at Penn State College's New Beaver Field, early 1950s

FIGURE 64 Cheerleaders fire up the crowd. Coeds joined the squad during World War II as a "temporary war measure," were ousted in 1949 at the urging of alumni, and returned thanks to student demand in 1951

FIGURE 65 An early Nittany Lion

FIGURE 66 Gene Wettstone, the first man to wear a mountain lion–style suit

FIGURE 67 Student honorary or "Hat" Societies, such as Skull and Bones and Parmi Nous, formed the honor lines through which players ran onto the field in the postwar era

Physical Education and Athletics Dean Carl Schott asked gymnastics coach Gene Wettstone to bring the Lion mascot back for pep rallies and games. Wettstone had a new mountain lion suit made, which he wore that fall while performing as the first of the modern Lions.

Coach Wettstone quickly recruited some of his gymnasts and then other students to carry on the mascot role. In the 1920s the mascot had led both the band onto the field and then the football team. During plays he would keep out of sight, but otherwise "cut up" with the cheerleaders, roar at the crowd, and accompany the song leader in conducting the alma mater.

Wettstone's inspiration in the 1940s was to have the Lion involved in ever more complex and outrageous "stunts." At a Penn game the Lion rode onto the field with a cheerleader dressed as William Penn on a bicycle built for two. At other games he arrived in a hot-air balloon, on a horse-drawn chariot, and, in a hard-to-top stunt, he emerged from an outhouse hastily constructed on the field by cheerleaders (and the Lion) disguised as workmen. There was a new stunt every game, and it was at this time that the Lion began doing push-ups equal to the points scored after each touchdown.

Coincidentally at this time, the campaign to move exuberant student celebrations from downtown onto campus resulted in the class of 1940 voting to fund the Nittany Lion Shrine as their class gift. At its 1942 unveiling ceremony, Joe Mason recounted the story of his creation of the Nittany Lion idea. The Board of Trustees in that same year designated the Nittany Lion as the college's official mascot, thus wedding statue and on-field mascot. In 1949 the man in the lion suit became formally associated with the cheerleading squad.

FIGURE 68 Heinz Warneke places finishing touches on the Lion Shrine

Off the field, the college's Public Information Department decided an improved press box was necessary when the grandstand was expanded in 1948–49. It was needed to accommodate both radio broadcasting and the print media, the electric scoreboard operator and the public address system announcer, as well as college officials and special guests. Walter Trainer and James H. Coogan, then in charge of sports publicity, visited numerous other colleges for ideas. The resulting four-level press box was deemed one of the best in the country by the Football Writers' Association of America.

For those who could not attend the game, the press fed the demand for news about the home team's triumphs and travails. Pennsylvania newspapers from Harrisburg, Pittsburgh, Philadelphia, Allentown, Lancaster, and Altoona, as well as local papers, AP and UPI wire services, and sometimes reporters from the *New York Times* and *Herald-Tribune* increasingly covered Penn State home games in the 1940s and '50s.

Radio coverage of the games originated with the WPAB college station between 1923 and 1932, run by the Electrical Engineering Department (the station was relicensed in 1926 as WPSC). There were no commercial sponsors; it was considered another means of college outreach at the time of President Thomas's Emergency Building Fund campaign. The first games broadcast were the 1923 Navy and Georgia Tech games, and they reached listeners all over Pennsylvania and neighboring states. Home radio sets had been largely a hobby before, but sales of commercially made sets took off in 1922 and by the mid-1920s every major college was broadcasting its own games. The Naval Academy even used radio to broadcast cheers and songs from Annapolis to Beaver Field, where they were played over the PA system.

By 1932 pressure from commercial stations had gradually driven college radio stations with few financial resources off the air. In 1938 the Atlantic Richfield oil company offered Penn State $1,500 to sponsor the entire schedule of games using KDKA in Pittsburgh, which originated a radio network with a few commercial stations. By comparison, Yale received $20,000 for the rights to its games. The first game on the new network was the October 1, 1938, Maryland game, with play-by-play by Jack Barry and color by Bill Sutherland. This network would eventually grow to over fifty stations, with noted Pittsburgh Pirates broadcaster Bob Prince handling the play-by-play in the 1950s and early '60s. At Penn State, radio profits in the 1930s and '40s were modest but the exposure for the college from radio broadcasts was more important.

Although there was consideration of televising the 1947 Navy game at Baltimore, the first televised Penn State game was a regional broadcast of the Boston University game in 1958, played in Boston, a 34–0 Penn State victory. Aside from bowl game broadcasts beginning in 1959, Penn State games were seen only occasionally on television network broadcasts. A few experiments with closed-circuit broadcasting were unsuccessful in the 1960s, although the University's own program reviewing the past game and interviewing coaches and various players began in 1965 as "Wednesday Night Quarterbacks" on WPSX-TV. While its name, format, and personnel have changed over the years, it continues today after each game and is broadcast over a variety of outlets.

State College Becomes a Destination

During one of President Sparks's many speeches, he joked that State College was "equally inaccessible from all parts of Pennsylvania." By World War I this was becoming less and less the case, even though regular passenger rail service directly to State College, or at least Lemont, had already reached its zenith in the early 1900s and begun to disappear. Notably, several special train trips to Penn State occurred after World War II. The last was the 1964 game against Pitt, with fans bused across campus from the station to the stadium and back.

Regular passenger service via the main line of the Pennsylvania Railroad was still available at Lewistown, about thirty miles away. During the 1930s the Lakes to the Sea Highway, later designated as US Route 322, was completed through State College, linking Penn Staters to the railroad connection with a two-lane concrete highway.

Bus service and, increasingly, private autos were available to some during the 1920s but were in full swing in the 1930s. By 1941 bus pickups and dropoffs at the Corner Room at College and Allen had ended, although the Boalsburg Autobus Company continued to offer school bus–type service from this location to the Lewistown train station at college break times. The Greyhound Bus Company built a station, with its classic Post House Restaurant, adjacent to the Bellefonte Central Railroad freight station. Although no longer exclusive to Greyhound buses today, the old freight station has been remodeled and still serves as the bus station, while the former Post House still houses a restaurant.

It seemed inevitable that air transportation to State College would eventually arrive. Sherm Lutz, State College's pioneer aviator, had established a landing strip and hangar between Boalsburg and Oak Hall in the 1930s. At first his business was confined to

flight instruction, excursions, advertising, and aerial photography. After World War II flight operations moved to his "State College Air Depot," a strip of land that is now a housing development, appropriately called "The Landings," between West College Avenue and Cato Industrial Park. There, limited air-passenger service under All American Airways (later Allegheny Airlines) was initiated in 1949. But the town was growing quickly toward the airport, and local citizens voted down permitting Lutz to put in a hard-surface runway so that planes could also land in rainy weather. As a result, Allegheny shifted its passenger service to Midstate Airport near Black Moshannon State Park.

With transportation options increasing, State College was becoming a destination for those who wanted to visit the college. Sports were an important reason for that travel, but not the only one. In addition to accommodating parents of students and large numbers of alumni for homecoming and the

FIGURE 69 Special excursion train bound from West Point to the 1961 Army game at Penn State, traveling on the Bellefonte Central Railroad tracks through the Bald Eagle Valley

FIGURE 70 Sherm Lutz's State College Air Depot; West College Avenue runs along the bottom of photo, the white farmhouse and three hangars in the middle constitute the depot's physical plant, and the air strip is the nearly horizontal clearing just above the farmhouse

FIGURE 71 The Autoport, ca. 1940

spring reunions, Penn State's extension activities drew increasing numbers of visitors. Building on the University's educational and research missions, people from all over the state and beyond increasingly visited the campus for professional meetings and conferences. The Nittany Lion Inn was created specifically to meet these needs while also catering to other campus visitors and those attending athletic events.

The State College Hotel, the lineal descendant of Jack's Road House (established in 1855), had grown to be a hotel with seventy guest rooms as well as a restaurant. In 1900 there was only a single restaurant in town, but by the 1930s there were a dozen, along with four hotel dining rooms. Just beyond the southern edge of town, the Autoport was created in 1936 as one of the state's earliest motels, catering to travelers with nine tourist cabins, a restaurant, and automobile service. Hotel accommodations in State College continued to slowly expand, but even in the 1950s there wasn't much development around the Autoport.

Perhaps an unrecognized example of these changes was the WPA state guide series. These guidebooks for most states and major cities, produced by the Depression-era Federal Writers' Project, included *Pennsylvania: A Guide to the Keystone State* (1940). College campuses and university towns like State College were included as places of interest to visit while following suggested tour routes. Attractions then were not all that different from today's; the Mineral Sciences Museum, the new Land Grant frescoes, and the college creamery were likely stops.

This is all to say that Penn State and State College were becoming easier to travel to and there were more accommodations for visitors in terms of hotel rooms and restaurants. Athletics was just one of the reasons to visit. However, the growing success of Penn State football in the 1940s and '50s could be seen in an expanded New Beaver Field and in ever-larger crowds with average seasonal attendance figures near or above the 27,720 capacity of the grandstands. With ten straight winning seasons through 1959, the football program was beginning to make a larger economic impact on the community as well as the college's athletic program.

BEAVER FIELD BECOMES BEAVER STADIUM

Following World War II, a massive social change began that would have a profound impact on America. Unlike previous postwar eras, the federal government created unprecedented new programs that would build a new middle class. One in particular revolutionized higher education by making it possible for vast numbers of veterans to attend college. Like many public colleges, Penn State saw enrollments rise dramatically and, with an expanded curriculum that responded to new funding opportunities as well as student demand, it would achieve a long-awaited transformation to university status.

The physical plant of the institution also changed to accommodate this growth. One of the most noticeable changes was the need to concentrate more academic buildings and student facilities in the campus core. As a result, athletic fields and grandstands for outdoor sports moved to the peripheries for both intercollegiate and intramural competition. The most dramatic of those moves relocated Penn State's football field to a new site for the second time in the history of the institution. However, New Beaver Field's steel grandstands had made it a far larger venue and a considerably more challenging move than that of a half-century earlier.

An Institution on the Brink of Dramatic Change, 1945–1960

The 1950s at Penn State were marked by the college's emergence from the disruptions of World War II and the turbulence of the postwar era brought on by the passage of the Servicemen's Readjustment Act of

FIGURE 73 President Milton S. Eisenhower

1944. The GI Bill, as it was better known, made it possible for every veteran to receive funding for college tuition and living expenses. Enrollments at University Park doubled between 1945 and 1950, rising to 11,132.

The student population grew so fast that a massive amount of temporary housing was required. This need was largely met with trailers, former military barracks, and other buildings left over from war-training facilities around the state. A temporary student union (nicknamed "The TUB" by students) was a converted USO building, and another temporary structure was used for classrooms. These units were erected on the east side of campus, most beyond Shortlidge Road, while new permanent dormitories were under construction.

The dramatic changes at Penn State included a new president. Milton S. Eisenhower, the past president of Kansas State College and the brother of General Dwight D. Eisenhower, took office in 1950. He brought a new energy to the college, and nothing symbolized that better than the long-overdue renaming of the school as the Pennsylvania State University in 1953.

Between 1950 and 1960 enrollments increased another 50 percent, to over 16,000 at University Park. New programs evolved to accommodate the growing student population, including the creation of a College of Business Administration, which soon ranked fifth in the number of majors among the University's nine colleges. Penn State's undergraduate centers around the state reopened (after being largely closed during World War II) and new centers were added, serving 4,605 students by 1960. Research also grew with the addition of the Garfield Thomas Water Tunnel, the primary scientific facility of the Ordnance Research Laboratory (renamed the Applied Research Laboratory in 1973), and the opening of the first functioning nuclear reactor on a college campus.

The 1950s also saw a resurgence of interest in intercollegiate athletics. As noted earlier, Penn State had sought to de-emphasize athletics starting in the 1930s in response to national trends condemning "professionalism" in college football. The "purity" policy eliminated athletic scholarships and restricted recruiting, scouting other teams, training-table meals,

and special tutoring for athletes. These changes led to the declining success of teams and forced a dramatic scaling back in the athletic caliber of schools Penn State played.

Alumni, however, never gave up hopes for a return to athletic respectability. The 1947 football team, made up largely of veterans, ended its season with no losses and a tie with powerful Southern Methodist University in the 1948 Cotton Bowl, Penn State's first postseason game since the 1923 Rose Bowl. However, policy changes were necessary if this success was to continue.

Along with the new president, Milton Eisenhower, in 1950 came both a new football coach, Brown University's Charles "Rip" Engle, and a new hope for more competitive athletics. Eisenhower supported the reinstitution of scholarships that had been initiated in July 1948 and, under his new athletic director, Ernest B. McCoy from the University of Michigan, appointed in July 1952, a new approach was created that did not sacrifice high-quality academics for athletic success.

Institutional Planning Anticipates a Move

If 1945 to 1960 represented a period of unprecedented growth and development, the University's planners could already see a bigger future on the horizon. The first wave of the baby-boom generation would reach college age in the mid-1960s and increased government funding for higher education pointed to even higher enrollments. During the Milton S. Eisenhower administration (1950–56), the University established a planning apparatus to anticipate the changes that were likely to come, an activity that Eisenhower's successor, Eric A. Walker (1956–70), gave even more prominence.

A volume of "Long-Range Planning Studies" was released in 1954, which detailed population, enrollment, fiscal data, and estimated trends through 1970. It anticipated expanding population and increased college-going for Pennsylvania's youth. Penn State believed that it would enroll the largest percentage of the state's college students of any institution, and that its enrollment would reach 21,000 by 1970 (18,500 at

FIGURE 74 Ernest B. McCoy

FIGURE 75 Charles "Rip" Engle

FIGURE 76 President Eric A. Walker

University Park), a nearly 60 percent increase from 1950. Over the course of the 1950s and '60s, these studies were regularly revised ever upward.

President Walker's first major public planning document, "Penn State's Future—The Job and a Way to Do It," was approved by the Board of Trustees on January 1, 1958, having been in preparation for more than a year. Besides laying out the changes in Penn State's mission of instruction, research, and service needed to meet the new demographic realities, a space utilization study to support these new programs was also included. The location of New Beaver Field, along with its adjacent baseball and practice fields, was shown overlain by buildings for Education and Psychology, with more to come after 1970. The relocated football field and track were plotted for the area beyond a planned dormitory complex that would eventually be called East Halls,

with parking designated for the area to the south and west of the field.

In addition, the study pointed out that the State College Planning Commission was also considering how the University's future growth would impact the surrounding borough. Chief among its recommendations was the creation of a bypass rerouting US 322 to the east of town and campus. The map showed a location following the approximate line of what is now known as Porter Road, connecting to Fox Hollow Road, with an inner-loop road also suggested, what would be today's University Drive. Along with reconfiguring campus roads to reduce through-traffic, one anticipated outcome was that a relocated stadium would be "accessible to the peripheral highways (U.S. 322 by-pass and the inner loop), and where adequate parking and adjacent practice and recreation fields can be built."

Thus, it became increasingly obvious that central campus, with the need for more dormitories, classroom space, laboratories, and faculty offices, could no longer afford to spare the space occupied by New Beaver Field and the other outdoor sports facilities adjacent to Rec Hall and the Nittany Lion Inn. As early as 1952 there had been hints that these facilities might move to the farmlands on the eastern campus.

From the standpoint of Athletics, this was also a likely outcome. Dean Carl Schott had stated in a School of Physical Education and Athletics Executive Committee meeting on February 13, 1952, that the farmlands along the proposed University Drive would "house Beaver Field when and if moved." The concept remained under consideration in the following years.

Four years later, on November 10, 1956, Dean of the College of Physical Education and Director of Intercollegiate Athletics Ernest B. McCoy wrote an extended memo to Comptroller S. K. Hostetter, long

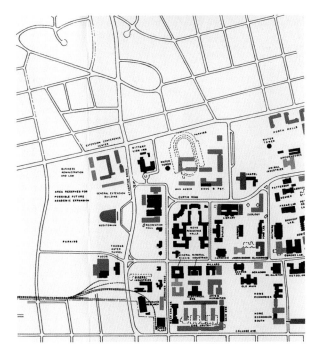

FIGURE 77 Map from 1958 long-range plan showing proposed Business Administration, Education, and Psychology Buildings occupying the Beaver Field site

Penn State's chief financial officer. McCoy outlined the problems with the current Beaver Field location. As enrollments increased, a larger percentage of seats was occupied by students, thus reducing gate receipts. Since football ticket revenue financed the entire athletics budget and supported additional recreation facilities for students, this was an untenable situation, and rising costs just made it worse.

The only alternatives seemed to be (1) cut other sports teams and raise ticket prices, which would both hurt students and potentially diminish income; (2) increase seating at Beaver Field, which would reduce adjacent space used by other sports and create less desirable, south end-zone seats for the public; or (3) "dismantle and move Beaver Field stands to a new location and build to anticipated needs." This would free up needed space for academics and allow for expansion of the stands without creating new problems. The fact that the stands were "of such construction that they may be reutilized almost one hundred percent for relocation in a new area" made the concept even more advantageous. The Board of Trustees' Executive Committee directed the administration to undertake a study of the future of Beaver Field.

Of course, there was precedent for this concept. In 1907 the first college plan included moving the sports fields from the center of campus to the northwest corner to make way for classrooms and laboratories. The decision to also move the existing grandstand to the new location, rather than to build a permanent stadium structure, helped make the project financially feasible. The wooden grandstand was easily disassembled, moved, reassembled, and later expanded as needed with additional wood construction.

We don't know if memories of the successful 1907–8 move contributed to the decision in 1959 to move and reuse the existing grandstands and support

facilities, rather than building an all-new stadium. Reusing the existing stands saved money and certainly made the decision a more attractive one. A campus plan done in 1930 that had suggested creating a new stadium at the west end of Pollock Road, on the southeast corner of the White Golf Course, never came to fruition. Similar crowding problems would have eventually occurred, and the stadium, if built of concrete, would have had to be demolished in order to move the field. Fortunately, the Great Depression and Penn State's economic woes actually perpetuated a grandstand design that would make an eventual move to the open fields of east campus farmland more feasible.

As noted above, the new Walker administration spent most of 1957 reorganizing its planning process and creating a new academic long-range plan. The "Penn State's Future" plan showed that moving Beaver Field was a major element in the space utilization plan. The University's Athletic Advisory Board, at its May 23, 1958, meeting, concluded, "It is now time to move the stadium and increase its capacity so that more desirable seats can be made available to alumni and other visitors, and to accommodate the anticipated student body."

In July 1958 the general guidelines for the move were in place, and the administration was authorized by the Board of Trustees to hire a consulting engineer to complete the study and make recommendations to the board for the move. Dean McCoy shared with one of his senior administrators in October 1958 that other athletic facilities, including the baseball field and football practice fields, would also move to the east. In regard to Board of Trustees approval, he confided, "I have been assured by those who should be in the 'know' that this will pass with some opposition, but it will pass."

The Relocation Project

The study first proposed in December 1956 finally resulted in the Board of Trustees' approval on January 24, 1959, of a project called the "Relocation and Expansion of Beaver Field." Dean McCoy's public announcement came the following day, reporting the official acceptance of plans to move the existing New Beaver Field structure to the far reaches of the east campus and to enlarge its seating capacity from 27,720 to more than 43,000. In spite of the official name of the project, the new structure was envisioned to be more than merely an expanded Beaver Field; it would be named Beaver Stadium.

The engineering firm entrusted with the responsibility for the detailed layout, design, and oversight of the project was Michael Baker Jr., Inc. Michael Baker '36, the firm's president, was a civil engineering graduate who was named a Distinguished Alumnus of Penn State in 1958. This was a fairly ambitious undertaking because the existing Beaver Field structure could not be disassembled until the conclusion of the 1959 football season, and the new, enlarged facility had to be ready for the first game of the 1960 season.

Although transporting the existing grandstand to its new location had to wait until after the 1959 home football schedule, the site work for what was to become Beaver Stadium began when the contract was executed on March 4, 1959. It was necessary that the site be fully ready to receive the relocated Beaver Field structure, including the press box, when the move could commence.

The site work consisted of two components that would be tightly intertwined. The first was the construction of the infrastructure needed to support the expanded venue. Portions of this work had to be completed before the second part could begin, while other infrastructure work would continue throughout the duration of the project. The second—and most visible—component was the construction of the superstructure of steel framing to support the steel deck and wooden benches that would expand the seating capacity by more than 50 percent. Together these two parts would lay the groundwork for the relocation and expansion that would modernize and enhance the football experience for both players and fans and justify changing the name from "field" to "stadium."

The infrastructure work commenced in March 1959. This phase had six individual contractors, with the primary role played by Wilson-Benner, Inc., as general contractor, and included a multitude of individual projects. First the entire site had to be graded, with provision for adequate drainage and the establishment of the graded base of the football playing field itself and the running track that would encircle it. The edges of the track were lined with a concrete curbing. Then an array of concrete footings had to be constructed to support both the steel framing and decking for the new stands and the steel grandstand that was to be moved from the Beaver Field site. In addition, the footings for the press box to be moved

from Beaver Field, plus the foundation for the adjacent elevator that would provide access to it needed to be poured.

Some of the ancillary features of the new facility were conveniently constructed at this point, such as the locker rooms, restrooms, and concession stands that would eventually be located beneath the superstructure itself. And, of course, the infrastructure work also included the establishment of water-supply lines, sanitary and storm sewers, and the integration of the necessary utilities, including underground electric and telephone lines.

The superstructure work, the other component of the project to be completed before the Beaver Field grandstand could be moved, was done by the Pittsburgh–Des Moines Steel Company, which had been responsible for the construction of all the steel grandstands beginning in 1934. The major part of this work was the construction of the steel framework and decking to support the thirty rows of new seating on both the east and west sides, with an additional ten

rows on the west side in front of the press box. Of course, decking for the new portion of the stadium included the necessary hardware for the attachment of new wooden benches for seating.

There was also construction of a walkway running the full length of the east and west stands that would be incorporated into the new construction where the new and moved sections connected (see fig. 85), and the switchback ramps beneath the structure to allow access to the new upper portion of the stadium. Additional steel framing had to be installed to increase the height of the press box.

When the new superstructure was in place, the construction site had the strange appearance of a stadium in which the first rows of seats were elevated some thirty feet above the ground. This portion of the structure had to be built in advance of the transport and erection of the Beaver Field structure that would form the lower part of the new stadium.

On November 14, 1959, the final game of the season was played against Holy Cross at New Beaver

FIGURE 79 Steel framework for west stands of Beaver Stadium

FIGURE 80 Rear view of upper portion of west stands

FIGURE 81 Dismantling of New Beaver Field

FIGURE 82 Beaver Field joins Beaver Stadium

FIGURE 83 Beaver Field decking repositioned
as part of Beaver Stadium

Field, and this cleared the way for Pittsburgh–Des Moines Steel Company to dismantle and relocate the existing structure. Beginning the week of November 16, one could hear the rat-a-tat-tat of the tools used to cut the bolts that connected the individual elements of the Beaver Field decking and seats.

The supporting steel framework was also disassembled, along with the press box and its supporting structure, and more than 700 pieces were then trucked to the new location at the east end of campus. Trees along Park Avenue had to be pruned to allow passage of the loaded trucks transporting the dismantled structure.

Once the elements of the dismembered Beaver Field structure arrived at the new location, they were systematically reassembled in front of the new upper part of the structure. The east and west stands were shifted northward twelve yards when compared with their position at Beaver Field to center them on the fifty-yard line of Beaver Stadium.

Incorporated into the relocated lower portion of the structure were two newly constructed wedge-shaped sections of seats on both the east and west sides to introduce a slight curvature to the stadium stands. This created a slight angle at the interface with the north-end horseshoe that required that the legs of the horseshoe from Beaver Field be shortened in order to compatibly fit the configuration of Beaver Stadium. This slight realignment provided better visual access to the playing field for those sections at the far reaches of the east and west stands.

After the relocated grandstand was in place, there were a number of necessary finishing touches: laying of the turf for the playing field; construction of the concrete pedestrian ramps outside of the running track to provide access to the lower-level seats; installation of a chest-high chain-link fence between the

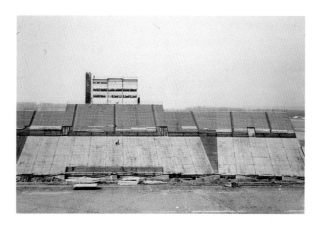

FIGURE 84 Wedge-shaped additions to Beaver Field stands

ramps and the track; and final surfacing of the running track. Exterior to the stadium itself, there was construction of the ticket booths, paving of the access walkways, and the final grading necessary for the grass parking areas adjacent to the stadium.

Beaver Stadium a Reality

Beaver Stadium was essentially an enlarged version of the former New Beaver Field—a large grandstand with a very simple supporting structure. The weight of the massive sections of steel decking was carried to the concrete footings by a forest of vertical steel columns spaced eighteen feet apart along the length of the stands and twenty feet apart in the transverse direction. The spacing between columns was somewhat smaller at the curved portion of the horseshoe. These columns were connected in both directions by horizontal pieces spaced at approximately twenty-foot intervals in the vertical direction. The columns and horizontal members thus formed two sets of planes composed of rectangular units running in intersecting directions—one set of planes along the longitudinal direction of the stands and the other

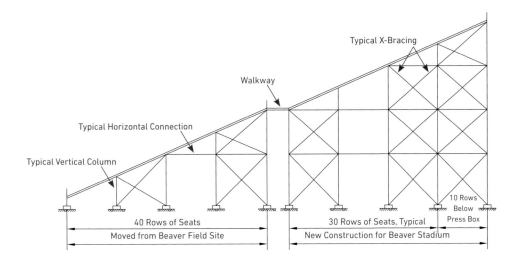

FIGURE 85 Typical cross section showing transverse plane for east and west stands. Drawing design by Harry H. West

FIGURE 86 First game at Beaver Stadium, September 17, 1960

perpendicular to them in the transverse direction. In the transverse direction, directly beneath the sloping stands, the rectangular units were reduced to quadrilaterals, with two nonparallel sides, or triangles (see fig. 85).

However, these rectangular units did not provide lateral resistance to wind loading or any sway motion caused by a boisterous crowd, so steel members were used to form a web of X-bracing across a number of the rectangular units within both longitudinal and transverse planes, thus adding stability

in both directions. The network of bracing was further strengthened by adding occasional rectangular frames, each of which possessed its own lateral resistance.

Beaver Stadium now provided seventy rows of seating on both the east and west sides, with the exception of the section in front of the press box on the west side where an additional ten rows of seating were included. The thirty rows of seating in the north-end horseshoe of Beaver Field were moved to the new location, but no additional rows were added there. The new facility thus had seating for 43,989 spectators, and with temporary bleachers at the open south end accommodating 2,295, the total capacity was 46,284. Figure 134, at the end of chapter 8, is a color-coded diagram in which items 6 and 7 show how the relocated New Beaver Field structure was integrated into new construction to form the initial configuration of Beaver Stadium.

A unique feature that was carried over from Beaver Field was an opening in the northwest corner of the structure to extend the track under the superstructure to enable straight runs for dashes, hurdles, and other events that could not include a curve. It later became a popular belief that the opening produced swirling winds on the field that would disrupt the passing game of visiting teams, the home team knowing how to cope with the phenomenon.

The first event in the new stadium was the 1960 spring commencement, which was held on June 11. The new structure was not completely finished, but it had reached the point where the commencement exercises could be held in keeping with the long-standing tradition of open-air ceremonies, which had been held at Beaver Field since 1939.

The first football game in the new stadium was a 20–0 victory over Boston University on September

17, 1960, before a crowd of 22,559. The maximum single-game attendance for the four home games of the 1960 season was 37,715 (approximately 10,000 more fans than could be seated in New Beaver Field), when Penn State's opponent was West Virginia. The expanded venue was clearly built for the future more than to meet the immediate demands of 1960.

The relocation was necessitated by a developing institution that required academic buildings, rather than athletic fields, in the central campus. At the same time, the increasing success of the football program resulted in increasing demand for seats from out-of-town fans. Combined with a growing student enrollment, and a venue that could not be effectively enlarged, there was little choice other than to transfer the athletic plant to the undeveloped eastern end of campus.

Thanks to the forced economy of not building a permanent, concrete stadium on the western side of campus, an amazing new athletic venue became possible. The move of the New Beaver Field grandstands and press box, and their reassembly as part of the expanded Beaver Stadium, was a truly remarkable engineering achievement that set the stage for the future growth of Penn State's football program and its home field.

"THE BIG LIFT"

Beaver Stadium's initial capacity, after the move from central campus, exceeded its needs but, by the late 1960s, both the permanent and the temporary wooden bleacher seats were regularly filled. With the increasing demand for tickets, seating capacity was increased incrementally. But by 1977–78 a substantial boost in seating numbers was necessary and it was accomplished with an unprecedented piece of engineering. Beaver Stadium's existing steel stands were lifted and twenty rows of new seats placed below them. The new seating was also extended to enclose the open south end of the stadium with forty more rows of seats. The lift project was an amazing accomplishment of design and construction—it had never been done to a stadium before and has never been done to one since. Like the move of the grandstand to its current location in 1959–60, it is a truly remarkable story.

Penn State and Football in the 1960s and 1970s

The mid-1960s through the 1970s is often represented by one phrase—the arrival of the "baby boomers," when the generation born between 1946 and 1964 began to reach college age. As mentioned in the previous chapter, Penn State enrollments, which had doubled between 1945 and 1950, now grew 50 percent more between 1950 and 1960, reaching almost 21,000, with 16,000 just at University Park. In the next decade, student numbers would double again at University Park to almost 33,000, while commonwealth campus enrollments increased almost 500 percent to over 26,000.

Naturally, faculty numbers, budgets, research allocations, and numbers of buildings also increased along with the expansion of the commonwealth campus system, as thirteen campuses were created,

FIGURE 87 Beaver Stadium as originally constructed, ca. 1968 (permanent seating capacity 43,989; temporary bleacher capacity 2,295)

FIGURE 88 Joe Paterno and Rip Engle

relocated, or upgraded to "campus" status in these two decades. From a cultural standpoint, the period from 1965 to about 1972 was the "sixties" of popular memory—times of debate and protest over minority and women's rights and enrollments, the Vietnam War, the University's defense-related research, and student lifestyles. The symbolic culmination of this time of activism came with student occupations of Old Main in 1969 and 1970, the latter turning into a violent confrontation between students and police.

President Eric A. Walker had already decided to retire on June 30, 1970. His successor, John W.

Oswald, had dealt with protests before as president of the University of Kentucky and then executive vice president for the University of California system. He spent his first year at Penn State in frequent meetings with students to maintain peace on campus. At the same time, enrollments continued to grow, but the University now struggled with stagnant state appropriations caused by inflationary pressures and economic recession.

The sixties and seventies demonstrated that Penn State had become the "big-time" school that alumni and students had anticipated in 1950. Coach Rip Engle was the "big-time" coach who would meet that challenge. Engle started modestly in his first two seasons, continuing the performance of Joe Bedenk's 1949 Lions but, nevertheless, Penn State still recorded twenty-six straight winning seasons under Bob Higgins, Bedenk, and Engle between 1939 and 1965. Most notably, the Lions had a 46–17 record over the 1959–64 seasons, winning three of four straight bowl games over Alabama, Oregon, and Georgia Tech. Penn State finished each of those six seasons ranked in the top 20, something that had only happened once since the vaunted 1947–48 teams.

In the words of football writer Ken Rappoport, Engle's approach was "low-key and curse-free." His introduction of the Wing-T offense in 1950 brought more passing into the Lions' game, and with the reinstated scholarships program Engle and his assistants were able to recruit increasingly talented players. They won key games against top teams, beating number one–ranked Illinois in 1954, and the previous national champion Ohio State Buckeyes in 1956. The Nittany Lions increased their number of intersectional games to two or three opponents annually, and they increasingly won more of those games than they lost. But, even with winning records, bowl game victories, and a number of All-American players, including Richie Lucas, Lenny Moore, Dave Robinson, Ted Kwalick, and Mike Reid, Penn State, as an Eastern independent, was still not considered the equal of the nation's great teams of the 1950s and '60s.

Success guaranteed that attendance at Penn State home games continued to climb. However, the new Beaver Stadium did not immediately yield sellouts. None of the 1960 games and only two in 1961 exceeded the permanent 43,989-seat capacity. Even for those two games, the bleacher seats were not full (bleacher seats were often opened to local youth for reduced ticket prices). But still, average home-game attendance remained steady through the early 1960s.

By 1965 Rip Engle was ready to retire. After fifteen years as an assistant and then associate coach, Joe Paterno became Engle's successor in 1966. With Paterno, Penn State had chosen a coach who knew the University and had absorbed the principles of recruiting top players who met Penn State admission standards and would graduate with solid academic records.

Paterno began to build a program that quickly challenged for national leadership. In 1967, his second season as head coach, the Lions amassed an 8–2 regular-season record, tied Florida State in its first bowl game since 1962, and were ranked tenth in the AP poll. There followed undefeated seasons in 1968, 1969, and 1973, plus one-loss seasons in 1971 and 1977. Overall, Penn State stood at 107–19–1 from 1967 through 1977, an 84 percent winning record. In those eleven seasons, the Lions earned ten bowl bids, won six of those games, and were ranked in the top ten in the AP poll nine times.

As a result of these continuing triumphs, attendance at Penn State home games jumped significantly. In the years 1967–69, when Penn State went 30–2–1,

average attendance began to climb, surpassing 50,000 in 1971. Even after a stadium addition in 1972, three of the first six home games were sellouts, and in the following seasons, each increase in seating capacity was matched by larger average game attendance. It is hardly surprising that, with Penn State's extraordinary record of winning seasons in the 1970s, demand for tickets continued the push for stadium expansion.

Incremental Changes, 1969–1977

The first step toward increasing Beaver Stadium's seating capacity came in November 1968 when the Board of Trustees considered final plans for the much-needed expansion of the press box and an addition of approximately 2,000 seats at the top of the west stands. They allocated $275,000 and $150,000, respectively, for the two projects, both of which were approved as designed by the Pittsburgh–Des Moines Steel Company. This work was completed in time for the 1969 season. The west stands were squared off by adding ten rows of seats on both sides of the press

FIGURE 89 Beaver Stadium with added rows of seats on west side and enlarged press box, 1969 (permanent seating capacity 46,049; temporary bleacher capacity 2,295)

box to match the ten extra rows that were provided in that section when the stadium was originally built. This increased the west side to eighty rows, adding 2,060 seats and bringing the total permanent seating capacity to 46,049; with temporary bleacher seating, the total capacity was 48,344. The expansion of the three upper levels of the press box more than doubled its length along the top of the west stands.

Following the 1970 season, the Board of Trustees recommended that bids be obtained for the addition of "approximately 9,230 additional seats" on the east side and north end of the stadium in September 1971. The cost was not to exceed $1.5 million, which included additional restrooms and concession stands. These additions were in place for the 1972 season and provided twenty more rows of seats on the east side and ten more rows on the horseshoe portion at the north end, increasing the seating by 9,194. This work was again done by the Pittsburgh–Des Moines Steel Company. Seating on the east side had now reached ninety rows, and on the north end, forty rows. Coupled with the eighty rows of seating on the

west side, the permanent seating capacity now totaled 55,243, and with the temporary bleachers, it reached 57,538. Figure 134 at the end of chapter 8 provides a color-coded diagram in which items 8 and 9 show these incremental changes to Beaver Stadium.

For both the 1969 and 1972 additions to Beaver Stadium, much of the design work for the expansion in seating was done by the University's Office of Physical Plant. Although there is no specific mention of it on the design drawings, the physical attributes of the Lambert Grandstand were carried forward in the new construction. Thus, the new portions of the stadium maintained a seamless interface with the old.

Beginning in the New Beaver Field "horseshoe" era, seating for special games was usually augmented by temporary bleachers at the south end of the field. Initially these bleachers were of wood construction, and they were removed after the football season to accommodate the use of the running track. Initially, these wooden bleachers were used at Beaver Stadium, but at some point a steel platform was constructed on which light metal bleachers were placed. The

FIGURE 90 Steel framework for the addition of twenty rows of seats on east side, 1972

FIGURE 91 Steel framework for the addition of ten rows of seats on the north-end horseshoe, 1972

FIGURE 92 Placement of steel decking for the addition of twenty rows of seats on the east side; steel framework for ten rows at north-end horseshoe, 1972

advantage of this arrangement was that the platform was high enough that the running track could be used without removing the bleachers. Finally, in 1976, these bleachers were replaced with semipermanent steel bleachers. This change more than doubled the bleacher-seating capacity to 4,960, which made for a total seating capacity of 60,203. Beaver Stadium now was recognized as the largest all-steel stadium in the United States (see fig. 134, item 10).

These bleachers were also positioned over the running track. However, because the overall structure and the foundation system were more substantial than for the previous arrangement, they permanently blocked the track. Because of this, Beaver Stadium could no longer serve as a venue for track and field meets, and these events were shifted to the new track and field facility south of Beaver Stadium.

An interesting item evident in the photo of the stadium with the semipermanent bleachers (fig. 93) is the contrast in the color of the seating. The brown color represents the painted wooden bench seats, most visible on the west side, which were originally used in New Beaver Field and relocated to Beaver Stadium. On the other hand, the ten rows of seats added in 1969 adjacent to the press box on the west side, and the twenty rows on the east side and ten rows on the north end that were added in 1972, were aluminum and therefore appear as gray in the photograph. During the 1980s the wooden seats were gradually removed and replaced by aluminum. This, of course, produced an abundance of 2×10-inch wood planks that were reused in a variety of applications both on campus and within the community. One adaptation can be seen from a bike trail that passes through the backyards of a residential area of State College, where there stands a crude lawn- and garden-equipment shed with brown, numbered

FIGURE 93 Beaver Stadium with semipermanent steel bleachers, 1976 (permanent seating capacity 55,243; bleacher capacity 4,960)

FIGURE 94 Reuse of wood planks that once served as seats at Beaver Stadium

planks for roof rafters. These are clearly remnants of what once served as seating at Beaver Stadium.

The installation of semipermanent steel bleachers at the south end of the field before the 1976 season increased the total seating capacity of Beaver Stadium to 60,203. Unlike the earlier major additions in 1949 and 1960, where the increased seating capacity exceeded the number of ticket-buyers, this modest expansion in seating capacity in 1976 couldn't satisfy the demand for tickets; another significant expansion of the stadium was needed. Once again, Penn State athletics faced the same challenge as in the two previous major enlargements; the construction would have to be completed between the last home game of the 1977 season and the first home game of the 1978 season.

Raising the Stadium

Initially, a feasibility study, which considered eight alternatives for expanding the stadium, was authorized by the Board of Trustees and undertaken by the A. W. Lookup Company of Philadelphia in 1976. Several possibilities were abandoned because they would have significantly increased the external footprint of the stadium, pushing fans in the upper end-zone seats farther from the action, among other problematic consequences.

One alternative, which would have maintained the existing external footprint, would have involved lowering the level of the field through excavation to allow seats to be added at the front edge of the existing stands, but this would prove costly and could have presented serious drainage problems. The option eventually chosen maintained the stadium's existing external footprint, but also left the playing field at its present elevation.

This option was a completely unique solution to the problem. The entire stadium and the press box would be lifted vertically by 12.6 feet, permitting twenty rows of seats to be added to the front edge of the existing stands, which would be closer to the playing field. In addition, a forty-row section of seating would be added to close the south end, transforming the horseshoe-shaped stadium into a bowl. Twenty rows of the south-end stands would be continuous with the rows added in front of the heightened structure; therefore, the southward external footprint

would only be modestly extended by the additional twenty rows of seats above. By choosing this option, the new additions increased the seating capacity by about 16,000, to just over 76,000 seats. Since all of the new structure would be of concrete construction, Beaver Stadium would no longer be the nation's largest all-steel stadium.

The expansion project was approved by the Board of Trustees on June 17, 1977. The successful bidder was H. B. Alexander and Sons, Inc., of Harrisburg at a bid price of $4,338,000. In addition to the expanded seating, the project raised the height of the press box and its existing elevator by 12.6 feet; it also added handicapped-accessible restroom facilities at the ground level and four additional restroom/concession buildings at the intermediate level. Michael Baker, Inc., the same consulting engineering firm that had guided the 1959–60 move, prepared the specifications and plans for the project and oversaw the work.

Of all the transformations that Beaver Stadium has undergone, this change was the most ambitious. Like the 1959–60 move from New Beaver Field, the schedule allowed only a nine-month window to accomplish this enormous undertaking, and a harsh winter added to the challenge. Clarence Knudsen, director of structural services for Baker, who was also on hand when the New Beaver Field grandstand was dismantled and relocated to become part of Beaver Stadium in 1959, said that he believed this was the "first time that a stadium had ever been lifted this way."

Two models (fig. 95) were prepared to show the pre- and post-configurations of the stadium. The post-configuration model clearly shows the effect of the lifting procedure and how the addition of the new concrete structure would produce the expansion of the stadium. The model also shows two additional possible expansions, one on the south end,

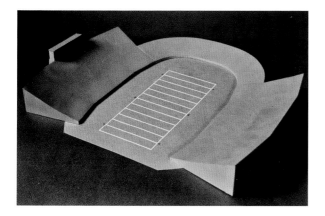

FIGURE 95 (LEFT) Pre- and (RIGHT) post-configurations of Beaver Stadium expansion of 1977–78. The post-view shows the lifted stadium, the twenty rows of seats over the track area (white), and the additional twenty rows on the south end (first arced layer beyond the white portion at the south end, which made a total of forty new rows in the south end). It also shows an additional twenty rows that were added at the south end in 1980 (second arced layer beyond the white portion) and full closure of the north end (never constructed)

FIGURE 96 Typical cross section showing the 1977–78 lifting strategy for the expansion of Beaver Stadium. Drawing design by Harry H. West

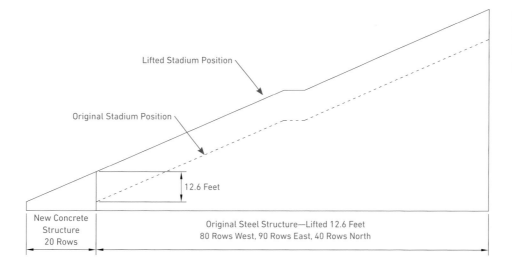

Lifted Stadium Position

Original Stadium Position

12.6 Feet

New Concrete Structure 20 Rows

Original Steel Structure—Lifted 12.6 Feet
80 Rows West, 90 Rows East, 40 Rows North

which would be carried out in 1980, and one on the north end, later abandoned. Robert Scannell, dean of the College of Health, Physical Education, and Recreation at the time, said that if the bowl were to be completed by these two additional expansions, there would be 120,000 seats. A sketch (fig. 96) shows the details of the change for a typical cross section of the stadium.

The extraordinary procedure for lifting the stadium required that the entire structure be divided into ten sections by cutting through the steel decking,

which was the base on which the seats were mounted. One section at a time was then lifted using hydraulic jacks, proceeding section by section around the stadium. The individual sections varied in length and included the full number of rows for that portion of the stadium. It would take forty jacks to lift the largest section, which weighed 500 tons. After all the columns for a given section were unbolted from their respective footings and the jacks were in place, the entire section was lifted to the desired height and new steel columns were then inserted to fill the gaps.

FIGURE 97 Elements of lifting operation: steel angle sections (LEFT); angles welded on stadium columns (CENTER); vertical legs of lifting frame (RIGHT)

  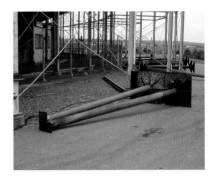

However, this is just a simple description of a much more complex process.

The subcontractor for the lifting phase of the work was MATX, Inc., a specialty contactor from Carlisle, Pennsylvania. A visit to the construction site shortly after work commenced in November 1977 would have revealed that MATX was gathering the materials needed for the lifting operation. Figure 97a shows several stacks of short pieces of steel angle sections that would be welded on opposite sides of every column of the stadium upon which an upward force would be applied to lift the structure (fig. 97b). MATX had also distributed beneath the stands pairs of approximately twenty-foot-long circular steel sections attached to large square steel plates (fig. 97c). These would be the vertical legs of a "lifting frame" (fig. 98) which was the key element in the lifting procedure.

A lifting frame was positioned between two stadium columns (as shown in fig. 98 and the detailed drawing of fig. 99). The two legs of the lifting frame stood vertically on the large square steel plates. They supported a horizontal "header beam" at the top of the frame, composed of two back-to-back "channels," on which the lifting jack was mounted. A lifting rod, which passed between the channels, connected the jack at its top to a bearing plate at its bottom. On

top of the bearing plate was a large horizontal steel "jacking beam," also composed of the back-to-back "channels." The vertical legs of the lifting frame and the lifting rod were sandwiched between these channels (section A-A in fig. 99) and, most critically, the jacking beam channels were in contact underneath the steel angles welded to each column (section B-B in fig. 99). All the lifting frames under a section were necessarily of the same height; therefore, toward the front edge of the stands the lifting frame was too tall to fit under the stands, so a portion of the steel deck had to be removed through which the frame could be positioned (fig. 100).

As the hydraulic pumps engaged the lifting jack, it pulled the lifting rod up, which in turn lifted the jacking beam, thus imposing upward forces on the angles and lifting the stadium columns to which they were attached. A similar lifting jack was positioned between every set of two columns for the section being lifted. The jacks (potentially as many as forty, depending on the size of the section) were hydraulically connected and equipped with safety devices to hold the load in case a jack failed. The process called for the lift to take place in four-inch increments, after which a team of workers provided a visual inspection of each lifting frame.

FIGURE 98 Lifting frame for column pair

FIGURE 100 Lifting arrangement for those cases where the frame was too tall to fit under the stands

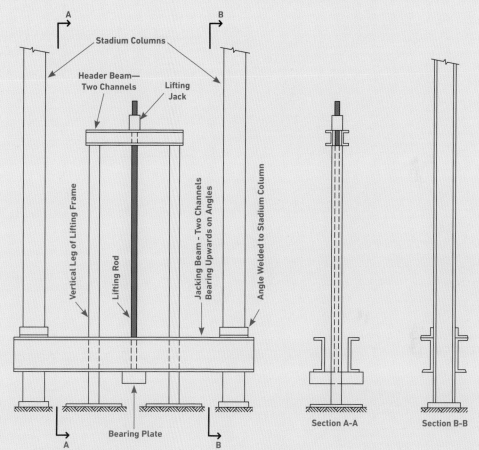

Section A-A   Section B-B

FIGURE 99 Sketch of lifting frame. The portion of the figure at the left shows the frontal view of the lifting frame. Sections A-A and B-B are internal views, rotated 90 degrees from the frontal view. Section A-A shows the view as seen from the A-A vantage point. The left, vertical leg of the lifting frame is sandwiched between the two lower-level channels that form the "jacking beam," which run between two stadium columns. At the top of the vertical leg the two upper-level channels are attached; these form the "header beam" that bridges the tops of the two vertical legs of the lifting frame. Most of the lifting rod, which connects the hydraulic lifting jack to the bearing plate, and part of the bearing plate are shown as dotted lines because, viewed from A-A, they are positioned behind the frame leg. The bearing plate places an upward force on the jacking beam as the jack pulls upward on the lifting rod. Section B-B shows the view as seen from the B-B vantage point. The jacking beam, composed of the two lower-level channels, bears upward on the angles that are welded to the stadium columns, thus lifting them the required 12.6 feet. Drawing design by Harry H. West

FIGURE 101 Lifting operations: hydraulic
pump (LEFT) and section lift (RIGHT)

FIGURE 102 Insertion of 12.6-foot column
under lifted section of stadium

FIGURE 103 Lifting snow-covered section of
east stands

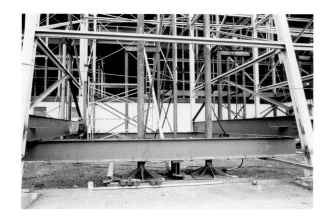 

Once the columns were lifted more than 13 feet, new columns were inserted into the gap and secured to the footing on the ground. The lifted structure was then lowered to 12.6 feet, the new bottom columns connected to the existing structure above it, and new horizontal beams and cross-bracing rods attached to stabilize the entire grandstand section.

The first section to be lifted was the northeast corner of the forty-row horseshoe section. This section was probably selected as a test run because it was only forty rows deep and of modest length. Work then progressed in both directions where larger sections were lifted. As time progressed and winter weather set in, the lifts became more challenging. The single lift (shown in fig. 103) includes two sections with ninety rows of seats on the east side. It was a massive task; additional snow weight required the stands to be cleared before the jacks could do their work. For the northwest corner of the horseshoe, in addition to the lift, new seating was added to cover over the opening that had existed for track events.

The same overall scheme was used to lift the press box. In this case, because the rear-facing columns were inclined, new footings had to be established. All of the lifts occurred on schedule except for the press-box lift, where wind conditions caused two delays. In one case, the lift had begun, but the press box had to be lowered because of the wind.

Completing the Addition

Simultaneous with the lifting of the steel structure was the construction of the new concrete portions of the stadium that added twenty rows of seating in front of the existing stands and forty rows of seats at the south end. Weather conditions again presented a challenge for this part of the work.

The first step was to establish a web of footings to support the structure. These footings were "drilled caissons," which were constructed by drilling a circular hole in the ground, inserting a cage of reinforcing steel, and filling the hole with concrete. The depth of each caisson depended upon the foundation conditions. The reinforcing steel of the cage protruded from the top of the caisson to become an integral part of the column that it would support. The top of the footing also had a keyed depression to ensure continuity with the column. A total of 273 eighteen-inch

circular reinforced concrete columns were cast in place atop the footings using disposable forms.

For the twenty-row sections on the east and west sides and the north end of the stadium, precast inclined beams were placed over protruding pins at the column tops which, in turn, supported precast concrete decking.

For the forty-row section at the south end, the lower twenty rows were constructed in the fashion as described above, but the upper rows required a taller supporting structure with more than a single story of columns. Here, the lower-level columns were connected together with precast concrete beams to form a framework, and the beam-to-column connections were completed with cast-in-place concrete to enclose the projecting reinforcing steel at the column tops and at the ends of the beam elements (fig. 107). This framework supported the inclined concrete beams, which were again placed over protruding steel pins at the top of the upper-level columns. In all, 1,160 pieces of concrete decking were placed, on which

the aluminum seating was attached. All the precast elements were constructed in a factory-type setting, trucked to the site and assembled like the pieces of an Erector Set. Separate concrete work was completed for the portals and ramps to provide access from the ground level under the stadium to the new concrete-based seating (fig. 108).

When the expansion was complete, there were 110 rows of seats on the east side, 100 on the west side, 60 at the north end, and 40 at the south end. The seating capacity was 76,017, and the crowds exceeded this number in all seven games of the 1978 season. Figure 134, at the end of chapter 8, is a color-coded diagram in which item 11 displays how the lifting project changed Beaver Stadium. With Penn State continuing its run of winning seasons, the stadium size still did not match the interest in Penn State football and the desire for tickets.

Evidence of the process of lifting the stadium in 1977 can still be seen today by game patrons wandering through the ground-level forest of steel columns.

FIGURE 107 Columns and precast beams

FIGURE 108 Portal construction and
supporting columns in place (note pins at
the column tops)

FIGURE 109 Placement of south decking

FIGURE 110 Finished expansion

FIGURE III   Angles on columns, evidence of
stadium lift

Patrons who look up, slightly above the 12.6-foot-
column section that was inserted when the stadium
was lifted, can see the short steel angles welded on the
opposite flanges of every column (fig. 111). Further
evidence extending out from the concrete stands is
the edge of the running track along both the east and
west sides of the field. The surface has been painted
green, but these strips are all that remain inside the
bowl of what was once part of the multipurpose stadium that served the needs of track and field meets in
addition to those of football.

According to Ken Getz, the job superintendent on
the 1977–78 expansion project for H. B. Alexander and
Sons, Inc., there were several significant challenges
associated with the operation. The weather was problematic, especially when snow had to be removed for
the lifts; there was a strike on campus, during which
work on the stadium was halted; and the concrete
columns had to be cast in place with great precision
to accommodate the placement of the adjoining
members at their tops. In addition, the construction workers were not allowed to walk on or move
equipment across the playing field to prevent further
compression of the soil base and damage to the field
drainage. In looking back at the project, it's clear that
the timing was tight, but the contractors finished just
in time—the hardware was being installed on the
locker-room doors the evening before the first game.

Penn State football had achieved recognition as a
national power by 1977. It had amassed a skein of 39
straight non-losing seasons. In 18 seasons in Beaver
Stadium, coaches Rip Engle and Joe Paterno accumu-
lated 154 wins against 44 losses and 1 tie, a 77 percent
winning record. The Nittany Lions went 31 games
without a loss in the 1967–70 seasons. They won 8 of
13 bowl games and were ranked in the top twenty in
AP and UPI polls in 15 of 18 seasons, 8 of those years
in the top ten.

Not surprisingly, the demand for tickets to watch
Penn State football games increased dramatically,
along with student enrollment at University Park
(which doubled between 1960 and 1977) and Alumni
Association membership (which more than quadru-
pled in those same years). Given these attainments,
increasing the capacity of Beaver Stadium from
43,989 to 76,017 (a 73 percent increase in the number
of seats) in this same time span does not seem at all
unreasonable.

FIGURE 112 Finished expansion fully
occupied (seating capacity 76,017). Note
light green strips on east and west sides,
evidence of former running track

THE NORTH-END UPPER DECK

The lifting of Beaver Stadium to add seats and convert it into a true bowl in 1977–78 increased the stadium's seating capacity to approximately 76,000. Unwavering fan support for the Nittany Lions immediately filled all the new seats. In 1980, with the demand for tickets continuing unabated, Penn State Athletics extended the south-end grandstands. This still did not satisfy the desire of fans to be part of the Nittany Lions' home-game crowd, and demand for tickets grew stronger throughout the decade. A number of expansion plans were considered, and in 1990–91 a north-end upper deck was chosen as the best approach. This option was complex and, it turned out, its completion was even more dramatic.

Penn State and Football in the 1978–1991 Period

In retrospect, the 1980s represented an extraordinary decade for Penn State. While there had been unprecedented growth in enrollment and by most other measures since World War II, Penn State had finally achieved the status of a mature institution. Its educational impact across the Commonwealth was now unquestioned, and its academic achievements were bringing it recognition as a university of national stature and, in some fields, international renown.

The social conflicts over civil rights, Vietnam, and the women's movement remained in contention to one degree or another throughout President John W. Oswald's tenure (1970–83). By the mid-1970s, however, the higher education environment was overwhelmed by recession, high unemployment, high inflation, oil shortages, flagging productivity, and the beginnings of globalization. Pennsylvania, in particular, was suffering from declines in coal mining, the iron and steel industries, and textile and clothing manufacturing. Its industrial dominance had been shrinking for some time, but the decline now accelerated, while the service and high technology sectors of the economy were just beginning to evolve.

Pennsylvania's state government faced severe economic challenges that mirrored those on the national scene. Governors Milton J. Shapp (served 1971–79) and Richard Thornburgh (1979–87) focused much of their efforts on fiscal stability. But Penn State faced a series of budgetary problems through the 1970s and '80s. Combined with a diminishing population of high school students in the state and erratic state appropriations, the precarious economic situation still limited the University's growth.

Bryce Jordan succeeded John Oswald as president in 1983 and served until the end of the decade. He had spent ten years as president of the University of Texas at Dallas, and then as executive vice-chancellor for the University of Texas system from 1981 to 1983, before taking office at Penn State. The challenges he faced included expanding Penn State's academic reputation by becoming a top ten public research university, continuing to build diversity by increasing the representation of minorities and women, and addressing the funding constraints Penn State faced.

Enrollments continued to rise University-wide, growing from approximately 62,000 in 1983 to 75,000 in 1990. At the same time, state appropriations grew more slowly than the rate of inflation in the late 1970s and 1980s. State budgets were sometimes late, and

Penn State appropriations were consistently less than the University requested.

As a result, private fundraising became a necessity to both enhance programs and make up for shortfalls. Despite a certain degree of skepticism and concern over the difficulty of a major development project, the Campaign for Penn State (1984–90) raised $352 million. Coach Joe Paterno, one of three vice chairmen for the campaign, had just completed an 11–1 1985 season, although the team suffered a disappointing loss in the national championship Orange Bowl game against Oklahoma.

Penn State football under Joe Paterno had achieved national recognition and respect in the 1970s and '80s. There were especially great teams in 1973, when John Cappelletti won the Heisman Trophy, and in 1978, when Penn State went 11–0 in the regular season before losing to Alabama in the 1979 National Championship Sugar Bowl game. Four years later the team would return to the Sugar Bowl, this time to defeat Georgia and claim the national championship. Over the 1977 through 1990 seasons, Penn State was 128–37–2 (a 77 percent record of victories), with nine or more wins in eight out of fourteen seasons and twelve bowl games. The decade's capstone achievement was a 12–0 record in the 1986 season and a second national championship with a now-legendary victory over Miami in the Fiesta Bowl.

After Joe Paterno's team won its first national championship, he was invited to address the Board of Trustees. He spoke of football only in the sense of the opportunity that it provided: "I think we've got to start to put our energies together to make this a Number One institution by 1990." The challenge was to use this "magic time for Penn State" to raise funds for better faculty and students, better libraries, and a university that wouldn't be afraid to discuss "new and disturbing ideas." He knew that there were vast numbers of alumni and friends who saw something special in Penn State and were committed to creating a great institution that just happened to also have a great football team.

The "Grand Experiment" Paterno first articulated in the 1960s had its roots in the academic expectations set by Coach Rip Engle, Athletic Director Ernest McCoy, and other faculty and administrators in the 1950s. However, Paterno wanted for his players both 80–90 percent graduation rates and all-around success in their lives beyond athletics.

The 1980s demonstrated that Penn State football represented more than just wins on the field. If Alumni Association membership is a useful metric, the fact that it had exploded from 25,000 members in 1970 to 100,000 members in 1987 should be illustrative. For alumni, students, and fans alike, the excitement and enthusiasm for Penn State football was hard to contain, and demand for seats to be a part of this great adventure grew ever larger.

More Incremental Changes, 1980–1985

The lifting of the stadium between the 1977 and 1978 seasons increased the seating capacity to 76,017, but a record crowd of 77,154 turned out for the 1978 season-opener against Rutgers (a 26–10 victory for the Lions). In fact, the new seating capacity was exceeded in twelve of the fourteen home games of the 1978–79 seasons. This motivated stretching the stadium capacity a bit farther. The most evident possibility to add seating was shown in the model presented in chapter 7 (see fig. 95). In the post-jacking model, there were three separate arcs of twenty rows each shown for the south-end horseshoe of the structure. The first two,

FIGURE 113 Construction of supporting structure for the addition of twenty rows at the south-end horseshoe

FIGURE 114 South-end horseshoe at sixty rows to match mid-height level of east and west stands; note the discontinuity of some aisles at the merge point with the forty-row level

totaling forty rows, had been completed in the 1977–78 expansion. Now, in 1980, the third arc of twenty rows was added, bringing the south-end horseshoe up to the sixty-row, mid-height level of the east and west sides of the stadium.

This expansion followed the same pattern of concrete construction used when the south end was closed in 1977–78. The foundations were established with drilled caissons, and the supporting structure

FIGURE 115   Enlargement of the press box in progress

FIGURE 116   Beaver Stadium in 1980 (seating capacity 83,770). This aerial view illustrates the method of parking cars and buses around the stadium before nearby structures and paved parking lots were added in the 1990s and after

was formed by an assemblage of precast concrete elements. In the process, 7,753 new seats were added, bringing the total seating capacity to 83,770. Item 12 of Figure 134 (at the end of the chapter) shows how this expansion contributed to the growth of the stadium.

Simultaneous with the increase in seating capacity came the near doubling in length of the top three levels of the press box through extension by equal lengths in both the north and south directions. This change satisfied the needs of an enlarged press corps and expanded the relatively sparse VIP seating. In addition, two new elevators were installed for the press box. The shaft of the original elevator was afterward used primarily as a conduit for cables for technological needs. Also, the first level of the press box was enclosed over its entire length with an internal passageway leading to expanded restrooms at the extreme north and south ends.

Another change in Beaver Stadium that occurred in this time period was the installation of lighting for late afternoon or night games. Temporary lighting had been used for a number of games in the early 1980s, with the expense generally borne by the TV broadcasters. A US Supreme Court decision in June 1984 declared the NCAA TV policy to be a violation of the Sherman Antitrust Act, thus opening the way for universities like Penn State to negotiate their own television rights free of NCAA control. Soon afterwards, the Board of Trustees approved the expenditure of $585,000 for the installation of permanent lights.

After the 1980 expansion, it appeared that the demand for increased seating capacity had been satisfied. Following an 8–4 1979 season, only three of the six games were sold out in 1980, but in the 1981–84 seasons twenty-two of the twenty-five games were sellouts, with two of the other three close to full. There

FIGURE 117 South-end walkway after removal of two rows of seats

FIGURE 118 Cantilever supports extend beyond the railing for addition of north-end walkway

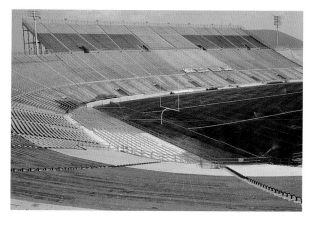

FIGURE 119 North-end, low-level seating

FIGURE 120 Corner ramp at the southwest corner

seemed to be no pressing need to expand the seating capacity at that point and, in 1985, the decision was made to address some issues that would enhance the game-day experience rather than to add more seats.

At the sixty-row, mid-height level of the east and west stands, there was a longitudinal walkway along which fans could move from entrance portals to either upper or lower rows. It was obvious that movement from section to section within the stadium would be enhanced if this pathway was extended around the north and south horseshoes to provide a circumferential walkway around the entire stadium. To establish this, two rows of seats were removed

from the top level of the south end to make room for the walkway, and a cantilevered support system was added to the north end to support its addition. The resulting continuous walkway greatly enhanced mobility for the fans, and it also made it possible for Blue Band members to move freely from section to section during the game and serenade the fans with Penn State fight songs.

To partially compensate for the loss of two rows of seating on the upper edge of the south end, new seating was established at the field level on the north end. This also served to provide additional seating for handicapped attendees. In the process, however, the total seating capacity was reduced by 400, rendering a new total seating capacity of 83,370. This was the first of only two times in the evolution of Beaver Stadium that the seating capacity was reduced. Item 13 of Figure 134 shows the effect of this modest reduction in the stadium's seating capacity.

Another change involved the construction of entry ramps at the four corners of the stadium. These additions greatly improved fan access to seating and concessions and, coupled with the circumferential walkway, they provided much improved movement throughout the stadium.

The North Deck

As Penn State's extraordinary successes on the field in the 1980s continued, especially after the 1986 national championship, the demand for tickets to attend home games surged. As noted earlier, one option for expansion entailed adding seats to the ends of the stadium, in effect creating a continuous bowl at 100 rows. There were several reasons why this concept was not adopted. First, additional seats at the top of either end of the stadium would have located fans at increasing distances from the playing field. Also, expansion at the south end encroached on Curtin Road, while an addition beyond the existing footprint to the north end pushed the structure into an area of lower elevation, which complicated the construction. Thus, while the complete bowl concept seemed a logical expansion at first glance, it simply did not hold up under scrutiny, given the practical considerations.

The next expansion concept to be considered involved the construction of an upper deck, which would have the advantage of adding seats that were closer to the field than the bowl strategy would have

provided. Attention was initially directed to the east side of the stadium. One possibility that was discussed involved constructing a multistoried building immediately behind the east stands, in which the lower levels would provide classroom and office space and the top levels would face the playing field and contain luxury suites. Such a building would have provided sufficient structural strength to anchor an upper deck that would cantilever out over the east stands without the need for supporting columns (much like what was eventually done for the south-end expansion in 2000–2001). Of course, this would have been a very expensive solution, and it was beyond the scope of consideration at the time.

A second possibility for an upper deck on the east side involved a simpler structural system, but it would have necessitated the inclusion of supporting columns at the forward edge. These columns would have passed through the lower-level stands, causing obstructions to the view of the playing field, which was judged to be unsatisfactory. Either of these scenarios would have yielded seats that were very high, because the lower level contained 110 rows of seats.

For these reasons, attention was again directed to the north end of the stadium with the idea of an upper deck over the horseshoe. Michael Baker, Inc. was selected as the engineering firm, and H. B. Alexander and Sons, Inc. was the successful bidder for the construction. The deck, completed in 1991, increased the seating by 10,033, bringing the stadium capacity to 93,967 at a cost of $12.1 million. This made Beaver Stadium the second largest campus stadium, surpassed only by the "Big House" at the University of Michigan, which had a seating capacity of 101,701. Item 14 of Figure 158 at the end of Chapter 9 shows how the inclusion of the north upper deck increased the seating capacity to 93,967. The scheme that was used here would have been more complex for the east side because of the height of the stands.

In order to cantilever an upper deck out over the existing horseshoe at the north end without supporting columns at the front edge, a substantial structure was required on which the deck could be anchored. The plan selected was to construct twelve large reinforced concrete bents (rectangular frames), spaced along the periphery of the north horseshoe, each radiating out perpendicular to the existing structure. The two columns for each bent were six feet square and sixty feet tall with a sixty-foot horizontal beam connecting the tops of the columns. As an integral part of these bents, corbels (projecting brackets) cantilevered out twenty-three feet to the rear to support the pedestrian ramp leading to the seats. These corbels were at differing heights from bent to bent in order to accommodate the slope of the ramps.

Anchored to the top of the bents was the steel structure that would support the seats and cantilever out over the existing low-level horseshoe. The deck for the seating area was steel, but it did not follow the pattern of the Lambert Grandstand of the earlier steel construction. It was welded steel plate material that was essentially custom designed for this application. The pedestrian ramps and concourse areas were reinforced concrete decks supported by steel beams with stay-in-place corrugated steel forms.

The plan produced a rather awkward-looking structure, but it was sound in concept. However, during the construction of the bents in the winter months of 1991, cracks developed in eight of the twelve bents near where the corbels and the columns met. Concern about the structural integrity of the ramp supports was an issue for all involved. What made it an even more critical situation, however, was that 10,033 spectators would be occupying

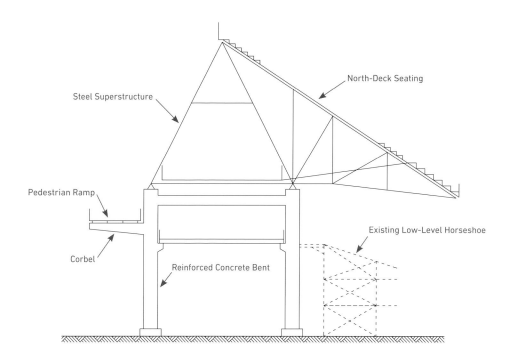

FIGURE 121 Structural arrangement for north-end upper deck. Drawing design by Harry H. West

FIGURE 122 Construction of bents and corbels

those north-deck seats (not to mention those sitting underneath the deck) at the upcoming season home opener against the University of Cincinnati on September 7, 1991.

The first step was to arrest the development of the cracks while the cause of the problem could be investigated. This was done by propping up each of

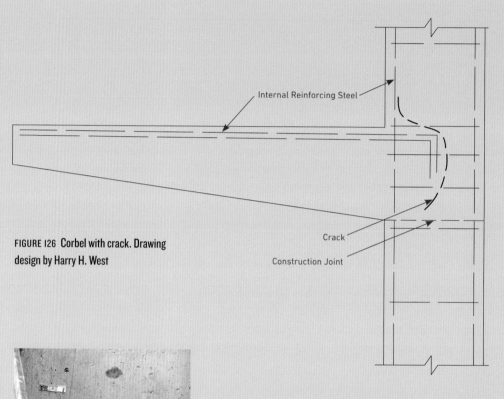

FIGURE 126 Corbel with crack. Drawing design by Harry H. West

Internal Reinforcing Steel

Crack

Construction Joint

FIGURE 127 Monitoring corbel crack

FIGURE 123 Typical corbel

FIGURE 124 Steel framing for upper deck and construction of pedestrian ramp

FIGURE 125 North-end upper deck

FIGURE 128 Temporary support for cracked corbel

FIGURE 129 Stirrup at the end of the corbel (LEFT) supporting cracked structure

the affected corbels with temporary modular structures built up from the ground. For some locations, openings were also cut in the concrete ramp so that steel cables could be passed through the openings to connect a stirrup, fitted over the bottom of the outer edge of the corbel, to the steel superstructure of the deck that was under construction above the bents.

Investigation of the problem suggested three possible causes for the cracks. One was that the forms supporting the bents and corbels had been removed too soon, which wouldn't have allowed sufficient time for the concrete to cure at the winter temperatures. The other two possible causes dealt with the reinforcing steel. Either the reinforcing steel in portions of the columns was inadequate or the reinforcing steel that provided the connection between the corbel and the column was not appropriately anchored. While the concrete-forms issue would have been a construction error, the other two would have been design issues that were internal to the already formed concrete; however, regardless of the cause, there was no way to redo the work to correct this problem.

Several corrective procedures were considered, but the one selected involved a post-tensioning procedure in which steel plates placed on the end of the corbel and interior face of the column would be pulled together by steel rods with sufficient force to close the cracks. The external steel rods would be tensioned to introduce axial compressive forces in the corbels, thus both closing the cracks and preventing any further ones. These rods were placed along the outer faces of each corbel, with their tension forces introduced by a jacking process against the steel anchor plates at the outer edge of the corbels and the inner face of the columns from which the corbels extended. While there were cracks in only eight of the corbels, all twelve corbels were post-tensioned as a precautionary measure. For those corbels at the highest level, it was not possible to place the anchor plate on the inside of the column. In these cases, the post-tensioning rods extended across the full width of the bent and were anchored at the outer face of the distant column. As a further precaution, the beams across the top of each bent were post-tensioned to prevent the possibility of the formation of cracks in these elements.

Once the post-tensioning was complete, a two-day set of tests was administered. A crane was used

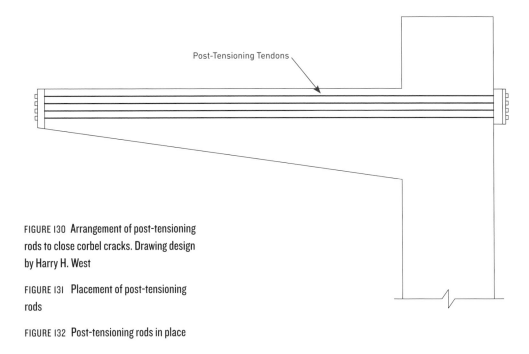

Post-Tensioning Tendons

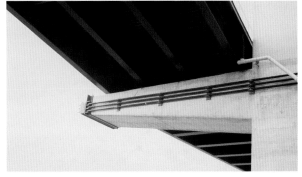

FIGURE 130 Arrangement of post-tensioning rods to close corbel cracks. Drawing design by Harry H. West

FIGURE 131 Placement of post-tensioning rods

FIGURE 132 Post-tensioning rods in place

to lift twenty-eight concrete road barriers into position on the pedestrian ramp, which the corbels supported. Each barrier weighed 7,000 pounds and all were left in place for twenty-four hours while the cracked areas were closely monitored. The barriers were then moved to a second location on the ramp and the procedure was repeated. The structure behaved well under the tests, and the deck was approved for use. During the Cincinnati game, the performance of the structure was monitored by taking measurements of the structural deflections under various loading conditions, including when the crowd did the "wave." The structure is flexible, so movement was discernible, but it was within the range of predictable and safe limits.

The retrofitted structure requires continuing observation. Concrete is a plastic material that deforms under sustained load; therefore, as the concrete responds to the compression introduced by the post-tensioning procedure, there is some loss of the tension forces in the rods. For this reason, the rods can be retensioned.

The addition of the north-end upper deck also introduced some new concerns that had to be addressed. Because the reinforced concrete bents that support the steel superstructure were not conductors of electricity, the new upper deck was not electrically grounded, as was the original all-steel portion of Beaver Stadium. For this reason, if lightning were to strike the steel superstructure, which

FIGURE 133 Beaver Stadium with north-end
upper deck (seating capacity 93,967)

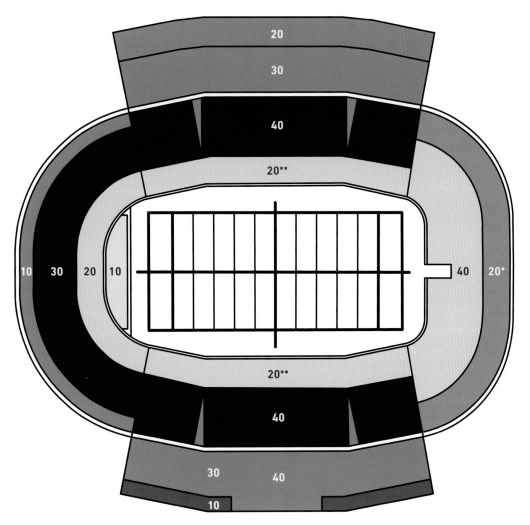

Section	Date	Seating Capacity	Color
6. Relocated New Beaver Field Structure	1959–60	27.720	⬛
7. Initial new construction of Beaver Stadium	1959–60	43,989[a]	⬛
8. Addition of 10 rows along portion of West Side	1969	46,049[a]	⬛
9. Addition of 20 rows on East Side, 10 rows at North End	1972	55,243[a]	⬛
10. Addition of semipermanent bleachers at South End; removed in 1978	1976	60,203	Not Shown
11. Lifting of Stadium to add 20 rows on East and West Sides and North End, and addition of 40-row Horseshoe at South End	1977–78	76,017	⬛
12. Addition of 20 rows at South End	1980	83,770	⬛
13. Removal of 2 rows at top level of South End for walkway, and addition of 10 rows of field-level seats at North End	1985	83,370	⬛

[a] Seating capacity augmented by 2,295 wood bleacher seats

FIGURE 134 Beaver Stadium development and expansion (numbers represent number of rows; see key for color code). Drawing design by Harry H. West

*South End reduced by two rows when top-level walkway was added

**East and West Sides reduced by three rows when seats for handicapped attendees were added at field level

was now, at 118 feet, the highest point of the stadium, it would be conducted down to the concrete bents and could cause damage. Therefore, lightning rods had to be installed along the top edge of the upper deck, with grounding cable, so as to conduct any lightning strike to the ground rather than to the concrete bents.

On a less technical note, the introduction of an upper deck created cover for an infestation of pigeons. This created a host of hygiene problems, the most troublesome being that the lower-level seats under the deck were frequently covered with pigeon droppings. Artificial owls were placed along some of the steel members with the hope of discouraging the presence of the pigeons, but the birds weren't fazed. Eventually netting was installed to prevent the birds from gaining access to the problem areas.

9

BEAVER STADIUM BECOMES A BUILDING

The most recent and striking expansion of Beaver Stadium added three levels of luxury boxes along the top of the east-side grandstands and two decks on the south end of the stadium, which were enclosed at the rear to become a building. Completed in 2001, this brought seating capacity to 106,537, making it still the second-largest football stadium on an American college campus.

After two national championship seasons in the 1980s, Penn State precipitated the realignment of the major intercollegiate athletic conferences by accepting an invitation to join the Big Ten in 1990. Soon thereafter, the football team won a conference championship in 1994, had a Rose Bowl victory, and narrowly missed winning a third national title despite its undefeated season.

The University's football fortunes in the 1990s, with a winning record of 97–26 and continuing demand for

tickets, led to this exceptional transformation. Beaver Stadium now met the needs of a university that ranked in the upper echelons of both American higher education and intercollegiate athletic programs.

Penn State and Football in 1991–2001

In the decade leading up to Beaver Stadium's most recent expansion, America's role in the world changed dramatically. With the end of the Cold War, the Middle East became the new focus of American foreign policy and security concerns. Following war in Kuwait, a succession of terrorist attacks through the 1990s led ultimately to the 9/11 attacks in 2001 and America's wars in Iraq and Afghanistan.

The 9/11 attacks also led to a heightened concern for security in the United States. Fear of random attacks in large public venues led to changes at Beaver Stadium in both access and the rules governing what spectators could bring into the stadium for games. In an era of political partisanship, economic crises, and persistent disharmony between various groups, the country's interest in sports as an outlet from the stresses of daily life remained strong.

Throughout the turmoil, Penn State retained its strength as an educational institution. At the University Park campus between 1990 and 2001, enrollments grew by 5 percent, to almost 41,000 students, after an increase of 15 percent in the 1980s.

However, Penn State had emerged from the 1970s and '80s with a significant backlog of physical plant needs. The Commonwealth's capital appropriations to Penn State had diminished acutely, leaving many gaps to fill. The Jordan era's phenomenally successful Campaign for Penn State had focused on student and academic programs, not "bricks and mortar."

One of the major accomplishments of President Joab Thomas's administration (1990–95) was breaking the logjam on state appropriations for new buildings. With Governor Robert Casey (1987–95), Penn State shaped a partnership combining private funds raised by the University with state appropriations for most new structures. General state appropriations for educational programs gradually began to rise through Casey's second term and the Tom Ridge (1995–2001) administration, reaching almost $300 million in 2000–2001—although still an actual 6 percent decrease in state funding from 1992 to 2001, after adjusting for inflation.

Nevertheless, the resumption of state aid for physical plant enabled the University to construct, add to, or renovate a substantial number of buildings at all Penn State locations in the 1990–2001 period. Soon after President Graham B. Spanier took office in 1995, he persuaded Governor Ridge to change funding arrangements for buildings. Instead of the state's choosing which projects to build from the University's physical plant priority list, Spanier was now able to select the projects the University believed were most needed and to secure a regular appropriation of state capital funds, which the University could match with privately raised funds, debt-financed funds, or auxiliary income in whatever combination fit.

One single event had a significant impact on some of this construction. On January 4, 1990, Penn State was invited to join the Big Ten athletic conference. In one of Bryce Jordan's final acts as president, Penn State quickly accepted the invitation. This came as a surprise to Big Ten coaches and athletic directors, and their opposition kept the question alive until it was decided by a final vote of the conference schools' presidents in June 1990. Penn State began conference play in football in the 1993 season, which ended just as ground was broken for the Bryce Jordan Center, the new home of men's and women's basketball along with many other nonsporting events. Joining the Big Ten did not mandate athletic facility upgrades, but Penn State realized that some of its venues did not match the quality of those of other members. Title IX also had an impact, as facilities for similar men's and women's teams required comparable improvements to venues, such as with baseball and softball.

In retrospect, the availability of parking had been one of the advantages of the 1960 move to Beaver Stadium. Aerial photos in earlier chapters (figs. 53 and 63) show the shortage of parking at New Beaver Field's west campus location compared to the open spaces surrounding Beaver Stadium. Forty years later, however, campus development began to surround the stadium and what were once pastures were now filling up with new construction.

The Bryce Jordan Center, the Intramural Building, the outdoor track and field facility, the Multi-Sport Facility and Ashenfelter Indoor Track Building, an improved Jeffrey Soccer Field, Medlar Field at Lubrano Park, Beard Field at the Nittany Lion Softball Park, the Pegula Ice Arena, new field complexes for lacrosse and field hockey, and an expansion of the football training facilities all formed an "athletic village" on the east campus. There were also the Physical Plant buildings, Fleet Operations, the Centre County Visitor and Convention Bureau, the Ag Arena, and the planned arboretum to the northwest, part of which had also been used for football parking. Now the parking lots in the fields, first to the south and later to the west of the stadium, were paved to provide parking for the Jordan Center and for commuter students and staff during the week, although at less density. Overall, the result of all the new construction and infill near the stadium was the reduction of adjacent football parking. Pushing both reserved and general parking even farther away from the stadium became a concern for marketing the football game experience.

Another area once planned for the vicinity of the stadium was the research park facility. Initially proposed in the Jordan administration, it was to include a conference center hotel, research buildings, and a business incubator. Innovation Park, as it later came to be called, was eventually located farther east, beyond the Mount Nittany Medical Center and the I-99/US-322 bypass, but football was no longer isolated from the rest of the campus as it had once been.

In 1991 Joe Paterno began his twenty-sixth season as head coach. Even though he would be sixty-five years old by the end of the season, Paterno showed little interest in retiring, despite periodic statements suggesting that his retirement would be four or five years in the future. The impressive record (102–33)

that he amassed over the years 1990–2000, along with six victories in ten bowl games and ten top-twenty rankings in the polls, amply demonstrated that JoePa still "had it." With planning for stadium expansion beginning in 1998, Penn State still "felt the magic" that Paterno had described to the trustees in 1983, and he welcomed the addition and what it could mean for Penn State football and the University.

Ironically, before football had moved out to the east end of campus in 1959, Rip Engle informed his staff of the plans. Then–assistant coach Paterno approached him after the meeting, saying, in effect, "Rip, I didn't want to say this in front of the other coaches, but you will be marked as the guy who ruined Penn State football." Paterno would eventually see the wisdom of the move. Recalling this conversation in a 2009 interview, Paterno said, "That was 80,000 seats ago. Shows you how smart I am."

Expansion Changes the Character of Beaver Stadium

Plans for the most comprehensive changes in the history of Beaver Stadium were first publicly announced in May 1998. They would essentially change it from a large grandstand into a building with a host of embellishments. The Board of Trustees approved the plan on September 10, 1999, at an estimated cost of $93 million. The transformation would include more than 11,000 new seats, broad pedestrian ways, improved spectator entrances, additional concession stands, new restrooms, elevators and escalators, new video scoreboards, sixty private luxury suites and larger VIP suites, a glass-enclosed club lounge adjoining 4,000 comfortable club seats, 6,000 general armchair seats on the upper deck, additional seats and improved access

for handicapped individuals, an All-Sports Museum and store, and a number of external enhancements to improve the appearance and access to the stadium. The architects for the project were HOK Sports of Kansas City and John C. Haas of State College; the structural engineer was Thornton-Tomasetti Engineers of New York City; and the construction was done under the joint management of Barton Malow/Alexander of Southfield, Michigan, and Harrisburg, respectively.

There was broad public support for expanding the stadium, although there were some objections to raising the height of the south end in a way that would eliminate the view of Mount Nittany. This sentiment was expressed through a series of letters to the editor of the *Centre Daily Times* and gave rise to a number of articles in the paper regarding the loss of the view. In addition, University Emeritus Trustee Helen Wise withheld her support of private boxes; she was quoted in the paper as saying, "I think it adds a touch of elitism that we don't need at a land-grant university." Nevertheless, as President Spanier pointed out, changes such as handicapped access, additional restrooms and concession stands, and improved accessibility and circulation benefitted all fans. Suite

and club seating revenue, in addition to increased ticket sales, would provide the major source of funds to make the stadium expansion project possible.

Although this expansion project would touch every part of Beaver Stadium, the most visible portions of the undertaking would be seen at the south end and the east side of the structure. With the addition of these two sections, the character of the stadium was dramatically changed. Even though the upper deck on the north end was structurally supported in a unique fashion compared with the previous parts of the structure, the nature of the venue remained that of a large grandstand; however, the changes underway through this phase of expansion would truly give Beaver Stadium a different character.

In addition to the dramatic alteration to the stadium environment, the structural changes were significant. The south-end addition is basically a multileveled building that wraps around the south horseshoe from which two seating decks cantilever out over the fifty-eight rows of the south portion of the bowl. The east-side addition is an elevated, three-story structure placed behind the 110 rows of grandstand seats. Both of these structures are composed of robust

FIGURE 136 Rising steel work for south end with precast concrete deck for seating area

FIGURE 137 Steel superstructure for east side

FIGURE 138 South end (LEFT) and east side (RIGHT) for Purdue game, September 30, 2000

structural steel frames. The decking for the seats at the south end was made up of precast concrete elements that were placed on inclined structural steel members that were a part of the steel superstructure. The other concrete decks were constructed with corrugated steel stay-in-place forms, supported by steel beams.

Unlike previous expansions of the stadium, these changes were more comprehensive than what could be accomplished between the end of one football season and the beginning of the next. In addition to the complexity of the construction project itself, there was the challenge to safely provide access for nearly 100,000 spectators in and out of a construction site for the 1999 and 2000 seasons. The actual seating capacity at this point was 93,967, but for those two seasons the attendance still ranged from 94,296 to 97,168. The Purdue game on September 30, 2000 (attended by 96,023 fans), was played before a background of looming steel construction on both the east side and the south end of the stadium (see fig. 138). Of course, there was no construction work on game days.

The south-end addition added 3,984 club seats in its lower deck, 6,584 general seats in its top deck, and 130 seats for handicapped individuals. But its major contribution in changing the nature of the stadium was the multilevel building from which these decks projected, and figure 141 gives the floor plans for the three primary, or enclosed, levels of the structure.

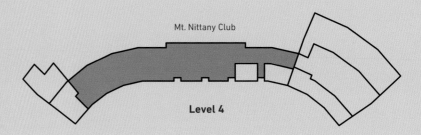

Mt. Nittany Club

Level 4

Letterman's Club

Recruiting Lounge

Level 2

All Sports Museum

Entrance Lobby

Home-Team Locker Room

Game Interview Room and All-Sports Museum Auditorium

Bookstore

Level 1

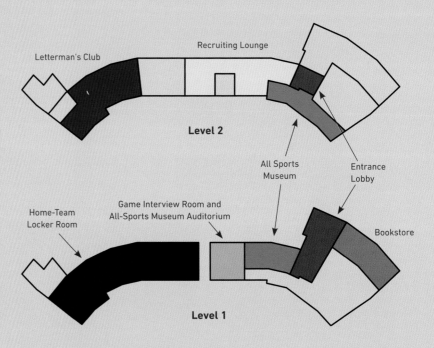

FIGURE 141 Floor plans for primary levels of the south-end addition. Drawing design by Harry H. West

FIGURE 139 South-end addition

FIGURE 140 Multilevel nature of south-end expansion

At level 1, or the field level, is the home-team locker room (including meeting, equipment, and training rooms). A plaque is mounted on the wall where the players enter the locker room; titled *With Pride from the Past*, it bears the following inscription reminding them of the unchanging Penn State uniforms:

When you put us on take a moment to think about those who have worn us in the past. Realize there have been student athletes for more than 100 years who have worn the same number you wear today. Know that in our long history we have not only endured the rigors of the field, but those of military conflict, financial depression, civil unrest and political deception. We are still here for you. Understand the players who wore us were ordinary young men like yourself, who went on to accomplish extraordinary things. They may have played at a different time but they had many of the same thoughts and feelings you have today. They played the game with a passion and desire that allowed them to accomplish what they could only dream of, prior to putting us on. We represent all the blood, sweat, tears, failures and successes of those who have played before. We are now entrusted to you.

"Wear us with pride"

Wear us, knowing that with simple things like hard work and perseverance you can achieve anything you can imagine.

"We are not magical, only cloth, you give us life"

Presented by the 1973 team, *Record 12–0*
Written by John Cappelletti,
1973 Heisman Award Winner

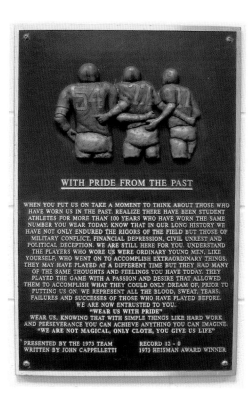

FIGURE 142 Cappelletti plaque at entrance to home-team locker room (LEFT); home-team locker room (BELOW)

FIGURE 143 Entrance to All-Sports Museum

Also at level 1 is the entrance to and portions of the All-Sports Museum, the press briefing room, and the bookstore.

The second level includes the remaining portions of the museum, a lounge for hosting high school recruits with an area that overlooks the press briefing gallery, several offices for stadium administration, and the Penn State Football Letterman's Club for former members of the football team.

Level 3 is a broad outdoor concourse that is continuous with the mid-level walkway that circumnavigates the entire stadium. At the edge of this concourse nearest the playing field, seating is provided for handicapped individuals. It is from the Curtin Road side of this level that fans gather to observe the arrival of the football team, which is transported from the Lasch Football Building on the familiar blue buses, and also the Blue Band, playing the fight songs as they march into the stadium.

The fourth level is devoted entirely to the glass-enclosed Mount Nittany Club, which is a lounge and dining area for members of the club. Its name is derived from the magnificent view afforded

of Mount Nittany from the elevated vantage point of the club. Within the lounge area is an array of concession stands that offer a wide variety of culinary options. Television sets are mounted throughout the club area so that nothing need be missed by those who retreat to the club for food or other comforts. Directly accessible from this level, through glass double doors, is the horseshoe of 3,984 comfortable club seats, which are supported by structural steel framework that cantilevers out over the mid-level concourse. As shown in figure 146, the club space has uses beyond game days.

The fifth level is the upper-level outdoor concourse, which provides access to another horseshoe of 6,584 armchair seats for non-club members. This level of seating is also supported by a structural steel framework, in which the lower portion cantilevers out over the upper reach of the club seating while the upper portion extends back to provide cover over the upper concourse. Mounted at the very top of the upper level of the south end is a video scoreboard.

At the southwest corner of the stadium is a wide, switchback, multilevel access ramp that rises above the bookstore, providing access for both game-day participants and small vehicles that are used for supplying the concession stands and for emergency purposes. At the southeast corner is a stairway that provides an alternate access path to the seating.

In addition to these means of entry, escalators at the southwest corner provide access to the mid-level concourse and the Mount Nittany Club, and elevators at both the southwest and the southeast corners accommodate handicapped patrons and provide access to the recruiting lounge, Letterman's Club, and Mount Nittany Club. These means of access replaced the smaller switchback ramps that were at

FIGURE 144 Penn State Football Letterman's Club

FIGURE 145 Recruiting lounge area

FIGURE 146 Mount Nittany Club, set up for a wedding reception

the south-end corners of the stadium. These had been constructed in 1985 when similar ramps were added at the north-end corners; they now sit tucked in under the north upper deck and continue in use.

The east-side addition is an elevated three-story building that extends the full length of the east stands, positioned behind the 110 rows of grandstand seats. The first and second levels each house thirty luxury boxes. The third level has several suites designated for

hosting dignitaries. Figure 151 shows the floor plans for each of the three levels of the suites.

The president's suite is centrally located at the top level and extends for nearly half the length of the structure. At the north end of the presidential suite is a connected legislative suite for government officials; both are used for entertaining donors and important guests. At the south end are several individual VIP suites for the athletic director, the head coach's

FIGURE 147 Wide switchback ramp at the southwest corner of the stadium provides access to all levels of seating. In addition, escalators behind the ramp carry patrons to the Mount Nittany Club level, and an elevator enables handicap access to all levels

FIGURE 148 On the southeast corner, a stair tower and elevator provide access to the Mount Nittany Club, the Letterman's Club, and all levels of seating

guests, the visiting athletic director, coaches' wives, and Nittany Lion Club members. In addition, at each end of the top level is an outdoor terrace to which individuals on this level can gain access as they wish. The combined seating for the suites is 1,960, and all three levels are served by elevators at the south end of the suites.

In all three levels, the corridor widens over the center section to create a glass-enclosed lounge that provides a magnificent view of the Nittany Valley facing eastward from the stadium. The width of this section of the corridor is greatest at the top level and diminishes some for the second and first levels as the glass enclosure tapers from top to bottom of the three stories.

Although the most evident portions of the expansion were at the south end and the east side, renovations to the rest of the stadium also had a major impact. A mid-level outer concourse was added beneath the east and west sides of the original stadium structure, which is at the same level as the concourse beneath the north-end upper deck and the lower concourse at the south end. It provides continuous wraparound access from point to point for the entire stadium. Along this concourse are new and improved restroom facilities and concession stands. At the concession stand locations, television sets are mounted to assure those who are purchasing food continuous viewing of any action on the field. All along the concourse, large images of Penn State star players of the past are posted, citing their accomplishments and the honors they received. The addition of this concourse required a strengthening of selected columns and foundations to sustain the additional weight. Although the space beneath the stands remained essentially unchanged, it was repaved and the spaces inside the four gates were dressed up.

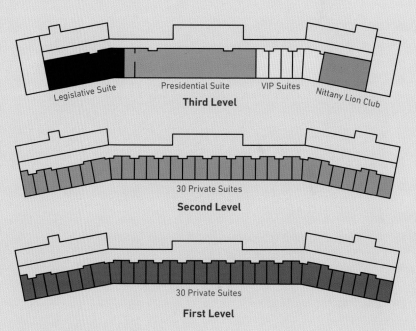

FIGURE 151 Floor plans for east-side suites.
Drawing design by Harry H. West

FIGURE 149 East-side addition

FIGURE 150 Three-level east-side suites

A video scoreboard was also to be placed atop the north end of the stadium, but, unlike the situation at the south end, the north upper deck was not designed to support its weight. Therefore, it was necessary to construct a tall supporting structure between the north deck and its access ramp on which the scoreboard was mounted. The structural framework of this scoreboard was fabricated on the ground and the assembled unit lifted into place. For this operation, a huge crane was brought onsite in August 2000 to accomplish the exceptionally high lift required to mount the 140-ton framework on its supporting structure. The uniqueness of the event, coupled with the unusual size of the crane, drew several thousand observers as preparations were made and the proper conditions for the lift were awaited. Once the framework was in place, all of the electronic work could be completed on the two scoreboards at the opposite ends of the stadium. In addition, narrow wraparound electronic display screens were mounted on the lower front edges of the north deck and the club seating deck to stimulate crowd participation, display scores of other games, and show major sponsor advertising.

In addition to the major enhancements to and renovations of the stadium, many exterior upgrades were completed. After September 11, 2001, security was significantly upgraded by the use of walls, fencing, and gates manned by security personnel. In the process, the exterior façade of the stadium was embellished, and broad pedestrian ways surrounding the stadium were created and tastefully landscaped. Banners were also attached to the lampposts to identify "Beaver Stadium, Home of the Nittany Lions."

When construction was completed in 2001, the stadium's seating capacity was 106,537, later stretching to 107,282 by converting some temporary seats to permanent. Then, after the 2012 season, three rows of seats were removed from the lower level of the east and west stands to provide additional handicapped seating. At the same time, permanent handicapped seating was added to the front of the lower-level seats at the north end. These changes added over 300 handicapped seats to the stadium, augmenting the more than 100 such seats that had already been introduced at the south end with the 2001 expansion. However, this reduced seating to 106,572, which was only the second time that the stadium's seating capacity was reduced. Figure 157 shows Beaver Stadium after the expansion completed in 2001. The color-coded diagram of figure 158 (end of this chapter) summarizes the stadium's growth through the addition of the north-end upper deck and the south-end and east-side expansions, with item 15 identifying the specifics after 2001.

For the first game in the newly renovated stadium, played against the University of Miami on September 1, 2001, the attendance was 109,313; the record crowd is 110,753 for a game against the University of Nebraska on September 14, 2002. Seating in excess

of the official capacity of the stadium is provided by small sections of bleacher-type stands that are scattered along the mid-level concourse, some of which are situated beneath the north-end upper deck. Many of these seats are reserved for the attendants who manage the parking lots before and after games.

Michigan Stadium, the "Big House" at the University of Michigan, built in 1927, had long been the largest stadium on a college campus. In 2007 its seating capacity was 107,501, making it slightly larger than Beaver Stadium, yet there was a brief period of time when Beaver Stadium would reign as the largest. A federal consent decree ended the US Justice Department's lawsuit against the University of Michigan claiming that it discriminated against disabled patrons by failing to provide adequate wheelchair seating, restrooms, and parking spaces. At that time, Michigan Stadium had only 81 seats for handicapped individuals, but under the terms of the decree, 96 additional seats were to be added for wheelchair users by the 2008 season. However, each wheelchair accommodation that was

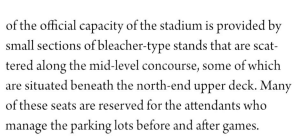

FIGURE 155 Hoisting the north-end scoreboard, August 2000

FIGURE 156 North-end scoreboard in place on top of supporting structure

FIGURE 157 Beaver Stadium (seating capacity 106,572)

added eliminated slightly more than 13 conventional seats; thus, the total seating capacity was reduced to 106,201, making the "Big House" second to Beaver Stadium. But Penn State would hold the title for only a brief time; after a major expansion completed in 2010, Michigan Stadium's total seating capacity was increased to 109,901, including 329 wheelchair seats, luxury boxes, and indoor and outdoor club seating, moving it to first place once again.

Even with the second-largest stadium on a college campus, Penn State fans continued their phenomenal support of Nittany Lion football. After a decade of success in the 1990s, expanding Beaver Stadium's capacity appeared to be the correct decision, as average stadium attendance continued at full capacity in the first decade of the 2000s. Changes in season ticketing policy in 2010 were expected to result in temporary decreases in attendance. However, the Sandusky prosecution and related legal issues beginning in November 2011, and the death of Coach Joe Paterno on January 22, 2012, launched a chain of events that could not have been imagined, let alone anticipated. Unquestionably, these factors affected attendance. Even more significantly, they produced the greatest challenges Penn State football had faced since the disruptions of World War II.

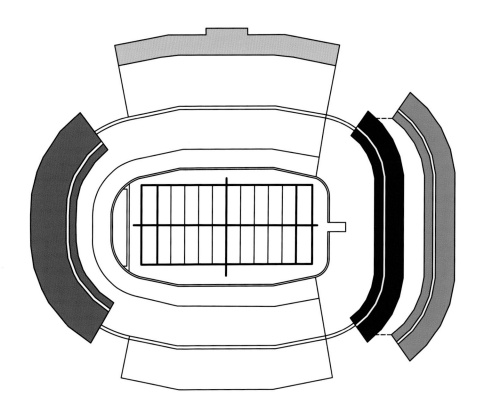

	Section	Date	Seating Capacity	Color
14.	Addition of North-End Upper Deck	1991	93,967	
15.	Addition of East-Side Suites	2001	106,537[a,b]	
	Addition of South-End Club Seats	2001		
	Addition of South-End Upper Deck	2001		

[a] Expanded to 107,282 in 2002 or 2003 when some temporary seats were made permanent
[b] Reduced to 106,572 in 2012 when handicapped seats were added at field level on East and West Sides

FIGURE 158 Upper-deck expansions and addition of suites for Beaver Stadium. Drawing design by Harry H. West

A NEW ERA

HERITAGE RECOGNIZED

The years since the 2001 expansion of Beaver Stadium encompassed the terms of presidents Graham Spanier (1995–2011), Rodney Erickson (2011–14), and Eric Barron (2014–). They have seen not only unprecedented transformations in Penn State football, but have also been the most dynamic and disruptive times in more than a century of Penn State history.

Penn State and Football Since 2001

Looking back over this period, it is easy to see that events on the world and national stage have had their effect on the University. Terrorism, war, and civil unrest have tested America's character and institutions. At the same time, mass shootings fueled further anxiety that became magnified on college campuses

with the 2007 events at Virginia Tech, where thirty-two died and seventeen were wounded.

Eleven years earlier, in 1996, Penn State had suffered its own campus shooting on the HUB lawn, when one student was killed and another wounded by a young woman with a history of mental problems. However, concern over campus safety has also prompted a heightened recognition of the need for protection from physical assault, sexual violence, racial harassment, and discrimination. Perhaps less disquieting, but still of concern to many students and their families, were continuously rising tuition, student loan debt, and an economy that struggled to provide the jobs that students hoped they would find upon graduation.

Despite these constant concerns, University Park enrollments between 2001 and 2015 increased

15 percent to almost 47,000, while total University enrollments grew to over 97,000. With the rise of the Internet and ever-expanding technology, much of the dramatic increase in total enrollment by 2015 came in the online course and degree environment of the University's new World Campus.

Penn State's overall budget continued to grow as well, reaching $4.9 billion in 2015–16, of which the largest portion, over 44 percent, went for general academic and administrative operations. Although state appropriations represented a relatively small but critical portion of the total budget, they fluctuated from a high of $304.5 million in 2010–11 under Governor Edward G. Rendell to a low of $203.4 million under Governor Thomas W. Corbett in 2011–12, precipitating a variety of stringent cutbacks. These changes reflected both the extraordinary volatility in the national economic picture and political conflicts that threatened the financial basis and viability of public higher education as never before.

Yet, through all the turmoil, Penn State achieved a number of dramatic and positive changes, including successful major development campaigns and nearly a billion dollars' worth of new construction and renovations across the Penn State system. The commonwealth campus system was reorganized so that students could earn a degree at every location. The Dickinson College of Law merged with Penn State, William A. and Joan L. Schreyer donated funds to create an honors college, new colleges of nursing and of information sciences and technology were created, and selected colleges, campuses, and departments were reconfigured.

Athletics overall saw extraordinary success, with nineteen NCAA team championships and fifty-six Big Ten team titles in nine women's and four men's sports. New athletics facilities, including Medlar Field at Lubrano Park, Beard Field at the Nittany Lion Softball Park, the Pegula Ice Arena, and new field complexes for lacrosse and field hockey were built while several other facilities were renovated between 2001 and 2015.

Penn State football, with its newly expanded 106,537-seat Beaver Stadium, entered 2001 with cautious optimism. Coach Joe Paterno was seventy-four years old as he began his thirty-sixth season as head coach. After a decade of outstanding seasons in the '90s, Penn State had a losing season in 2000, only the second since 1938. Still, the confidence of the fans remained strong and the newly expanded stadium sold out for the 2001 season. Although the 2001–4 campaigns produced only one winning season and a 21–26 record, attendance still remained at almost 98 percent of capacity.

Despite critics' talk of retirement and "Joe must go" websites, Paterno clearly had no interest in leaving, and the University signed him to a four-year contract extension in May 2004. While some were stunned, the seventy-seven-year-old coach had his eyes on the future. The 2005 team went 11–1 and was ranked number 3 in both national polls. It shared the Big Ten championship with Ohio State and Paterno was voted Coach of the Year for a fifth time. Over the 2005–10 seasons, Penn State amassed a 58–19 record, won four of six bowl games, shared another Big Ten title in 2008, and had five top-twenty-five rankings. The average attendance over those six seasons exceeded stadium capacity.

Despite the success on the field, things began to change in 2010. A new seat-licensing plan—the STEP, or Seat Transfer and Equity Program—was introduced. Under this program, a personal seat donation, along with the price of the ticket, was required for a season ticket in a particular stadium section. These donations ranged from $100 for the north end zone and seats in the north deck and upper south deck, to $2,000 for the comparatively small number of backed-armchair seats in prime locations between the forty-yard lines.

Those who wished to stay where they were would pay the personal seat donation for the section where their seats were located; those who wished to upgrade could move to a better section by increasing their donation, while those who downgraded would pay a reduced donation. Some simply chose to give up their season tickets.

Athletics described the changes as necessary to meet the funding needs of Penn State's extensive sports program. While the administration never released a number to account for those who gave up their tickets, average attendance dropped by about 3,000 a game in 2010, about 3,000 more in 2011, and by almost 5,000 more a game in 2012—more than 10,000 overall.

However, 2011 turned out to be an *annus horribilis.* The sixty-one-year career of Coach Joe Paterno came to an unimaginable and sudden end as a result of the November 5 arraignment of Jerry Sandusky on child sexual abuse charges. Paterno's removal from his head coaching position, along with the end of Graham Spanier's presidency, both by Board of Trustees action on November 9, precipitated a series of events that extended over the next several years and tested the future viability of Penn State's football program.

The University began to experience a firestorm of legal and leadership challenges that fed a scandal-driven media frenzy. In the months and years that followed, Penn State underwent a continuous transformation of its football and intercollegiate athletic program and participated in a larger conversation about the role athletics should play in the life of higher education and of Penn State in particular.

Yet, despite the chaos and controversy, the popularity of football persisted. The transitions in the football program following the January 2012 death of Coach Paterno continued, and the NCAA sanctions were largely overturned by early 2015. While the future course of football success is, as always, unknown, the interest in watching Penn State football and being part of the stadium experience remains, and average attendance increased over the 2014–15 seasons, under Coach James Franklin, to just over 100,000.

At the same time, Beaver Stadium as a venue is over fifty years old. In 2015 an Intercollegiate Athletics facilities master planning effort was launched. Early announcements mentioned that renovation or replacement of Beaver Stadium would be examined as all possible futures were evaluated. The inclusion of the word "replacement" launched a local media tumult and a storm of alumni angst. Whether inadvertent or not, for the first time in fifty-five years the public began to discuss what the future might hold for Beaver Stadium.

Regardless of how the stadium might change, the football game-day experience that takes place there is suffused with memory and nostalgia. While the very structure itself stirs deep feelings for Penn Staters, the University's football history is commemorated through both tangible reminders of that heritage and the game-day rituals and stories fans have come to cherish. Just as the stadium has changed frequently since 1960, some of those rituals and living traditions have evolved through the years as well.

History and Tradition Represented

Like the classic, plain uniforms the players wear, Penn State football for many represents tradition and heritage. Beaver Stadium plays a key role in that tradition as well, and there are numerous reminders of history at every Penn State home football game. First, there's

the name. The uninitiated are confused—if Penn Staters are the Nittany Lions, why is the stadium named after that industrious rodent, the beaver? If there's no loyal Penn Stater around to correct this mistaken impression, there are monuments and signs.

The earliest was a plaque mounted on a large sandstone block to honor Governor James A. Beaver, who secured the funding for Old Beaver Field. In 1954 it was placed next to the New Beaver Field ticket booths, near the Nittany Lion Shrine, as the gift of the Class of 1909 at its forty-fifth reunion. When the New Beaver Field grandstands were moved after the 1959 season, the monument was also relocated to outside the ticket booths at the southwest corner of Beaver Stadium. Following the 2001 stadium expansion it was moved again to its present location at the student gates on the southeast corner, where it serves as a permanent history lesson to new generations of students.

The bronze plaque features a bas-relief portrait of Beaver with a text that commemorates his service as governor, board president, and acting president

of the college during the period following President Atherton's death. As both an elected trustee and as governor, Beaver served on the board continuously for forty-one years. His popularity with students is also evident from the inscription, which names him "Beloved friend and counsellor of the Class of 1909."

However, those who miss the monument might see the blue and white Penn State historical marker that retells his story. Created under the auspices of the Penn State Alumni Association in 1989, it is one of seventy-five markers situated around the University Park campus and at ten other campuses that commemorate Penn State traditions, events, academic accomplishments, and notable individuals.

Two additional historical markers at the stadium are located along Curtin Road. One, entitled "Champions for Equality," is located to the southwest of the All-Sports Museum entrance and commemorates Penn State's role in breaking the color barrier in intercollegiate football. The second, halfway between the stadium and the Bryce Jordan Center, commemorates Penn State's entrance into the Big Ten in 1990 and the conference's historic role in the promotion of both athletic and academic excellence. Elsewhere on campus, there are markers at the original site of Old Beaver Field and at the Nittany Lion Shrine.

If the historical markers don't satisfy the desire for a deeper dive into Penn State's athletic heritage, the All-Sports Museum, which opened in 2002, will likely meet the need. The museum is located on two levels in the southwest corner of the stadium, with its entrance marked by a canopy extending out from the façade of the enclosed south end. Visitors to the 10,000-square-foot museum start their tour on the second level with an overview of Penn State athletic history and exhibits for men's and women's indoor sports. Returning to the first level, they see displays on outdoor sports and an array of Penn State's athletic awards and trophies. Immediately adjacent to the museum is the football media room, which as the Clemens Family Theatre doubles as space for special museum programs and films on nongame days.

The museum contains hundreds of photographs and artifacts, interactive video and information kiosks, and opportunities for some hands-on fun with a few of the less familiar pieces of sports equipment that Penn State athletes use. A branch of the Penn State Bookstore serves as the museum gift shop and is located adjacent to the museum.

Inside the stadium, additional special features remind fans of Penn State's football history and traditions. On the front of the east-side suites, a display of sixteen historic years honors the most memorable seasons that Penn State has achieved since it began playing intercollegiate football almost 120 years ago. The early years mark undefeated seasons, as the program evolved from victories over Pennsylvania schools to intersectional games and nationally ranked

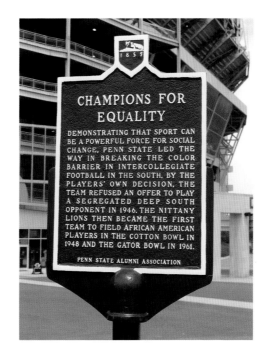 

opponents. The year 1968 is the first date for the Joe Paterno era, which included five undefeated seasons, two national championships, our first Big Ten championship and Rose Bowl victory. The addition of "2012" did not mark an undefeated season, but rather a gritty 8–4 record in Coach Bill O'Brien's first year, deemed to be one of Penn State's finest performances, defying the onerous and ultimately retracted NCAA sanctions.

While the chronology of milestone years reminds us of Penn State's long and distinguished football history, the victory bell represents a custom that is renewed with each home game. The tradition of ringing a victory bell to mark each point scored in a Lion win began in 1964. Coach Rip Engle took his players to the front of Wagner Building, near the stadium, to mark their victories by ringing the mounted bell of the USS *Pennsylvania*, a battleship that survived the attack on Pearl Harbor. In 1979 the Class of 1978 presented

FIGURE 161 Alumni Association's markers: Champions for Equality (LEFT), Big Ten (RIGHT)

an antique bell as a class gift. It initially hung on the top of the south end scoreboard and the Nittany Lion mascot rang the victory bell for each point scored in a winning game. With the stadium expansion in 1999–2001, the bell was moved to its present location next to the tunnel through which the players pass to go on and off the field. After the players and coaches sing the alma mater in front of the student sections, a tradition that began in 2012, they each ring the bell once to mark the victory as they depart the field.

Reflections of Heinz Warneke's Nittany Lion statue have also added a sense of history and tradition to Beaver Stadium since the 1999–2001 addition. Fans approaching the stadium along Curtin Road from the west can see the large metal Nittany Lion silhouette sculpture that fronts the massive ramp rising at the southwest corner of the stadium above the museum entrance and bookstore. The silhouette sculpture gazes back in time to Warneke's iconic Lion statue at the other end of Curtin Road, connecting the new

center of Penn State athletics at Beaver Stadium with the pre-1960 center at Rec Hall.

Another example of Lion imagery is the Nittany Lion weathervane. Perched atop the southwest corner of Beaver Stadium at 110 feet, this replica of Warneke's Lion in copper plate matches both the size and design of the limestone original. However, at 1,600 pounds and 10 feet 3 inches in length, he is likely the second-largest working weathervane in the world, as he was when erected in 2001. This Lion was crafted by metal sculptor Travis Tuck of Martha's Vineyard, Massachusetts. He was donated by Penn State alums Joel N. and Peggy Myers. Once the breeze reaches about ten miles per hour, the big cat pivots on his steel ball-bearing base and faces into the wind. This Nittany Lion was the topping-out gift to complete the 1999–2001 stadium expansion.

The move of the "Original Nittany Lion" to the All-Sports Museum provided a further historic link. The stuffed and mounted Pennsylvania mountain lion

specimen sat in Old Main's natural history museum in 1907 and perhaps inspired Joe Mason when he first imagined a Nittany mountain lion as the great and fearless symbol of Penn State. The lion was restored in 1993 and housed in a custom-built display in Pattee Library before the move to the stadium.

Two more Nittany Lions were added to the stadium in 2014. The familiar Lion logos, illuminated by 1,400 LED lights, can be seen for miles, day or night, looming above the massive superstructure of the stadium. The panels use the Intercollegiate Athletics Logo, which was created in 1983 after Penn State's first national championship, to meet the demand for Penn State clothing and memorabilia with a university emblem. Penn State is believed to be the first university to trademark and use a logo for commercial purposes.

On May 15, 2014, the first 35 × 25-foot panel, weighing in at 6,500 pounds, was assembled on the ground and lifted into place on the rear of the north scoreboard by two gigantic cranes. Less than two weeks later, with the temporary closure of Curtin Road, the second logo was installed on the south scoreboard. Similar logos are common features at other major university stadiums and were the final step of a larger project to renovate the massive scoreboards with new full-screen, high-definition video panels and an upgraded sound system.

Perhaps the most controversial monument at Beaver Stadium is one that is no longer there. The seven-foot tall, 900-pound statue of Coach Joe Paterno running onto the field, followed by four players in bas-relief on a wall behind him, flanked by two maxims in large, bold letters and a collection of tablets showing the game results of Penn State teams from 1966 to 2011, was installed in November 2001.

The statue of Coach Paterno was created by sculptor Angelo DiMaria of Reading, Pennsylvania. DiMaria, who had never seen a football game, never met Paterno and worked only from photographs as was his practice in sculptures. The statue and plaza were commissioned by friends and admirers of

FIGURE 163 Victory Bell

FIGURE 164 Nittany Lion silhouette on southwest ramp above the Penn State Bookstore

FIGURE 165 The Nittany Lion weathervane

FIGURE 166 Nittany Lion illuminated logo on rear of scoreboard

FIGURE 167 Upgraded scoreboard, 2014

Coach Paterno in anticipation of his becoming the NCAA Division I-A all-time victories leader with 324 wins. DiMaria called it "a tremendous monument. Although Joe Paterno didn't really want the statue up. He was very humble about it."

Through the years the illuminated courtyard became a favorite photo spot for visitors and new graduates in cap and gown. Although Paterno's unequaled success extending to 409 victories was a matter of pride for Penn State fans, the sculptor said the statue was not about wins. He considered it the personification of the values and integrity that Joe Paterno taught through both his coaching and his service to the University.

In the aftermath of the Sandusky scandal and threats to destroy the statue, the University made the controversial decision to remove both the statue and courtyard, which had become, in the words of some in the media, a "lightning rod" for the emotions generated by all the events of that challenging period. Ten days after the Freeh Report was issued and one day before the NCAA sanctions were announced in July 2012, the entire installation of statue, walls, and courtyard was dismantled and the space re-landscaped. Its absence has not erased the memory of Coach Paterno's accomplishments and the generosity and leadership of Joe and Sue Paterno. Signs, flowers, and memorabilia are placed at the site in his honor at every home game.

While there is much that remains unsettled after all these events, a number of other legacies remain, including the Paterno Library, the Paterno Family Professorship in literature, the Paterno Fellows program for liberal arts students, several scholarship and graduate fellowship funds, important financial assistance to the Department of Classics and Mediterranean Studies, the Richards Civil War Era Center, and the Pasquerilla Spiritual Center. One further piece of Penn State tradition also persists. The Berkey Creamery's Peachy Paterno ice cream still remains one of the favorites of Penn State students, alumni, and friends.

These are among the most notable memorials, symbols, and traditions that form the community memories of Beaver Stadium. According to historian Christopher Phelps, "When a university names a building after someone or erects a statue to that person, it bestows honor and legitimacy. The imprimatur of an institution of higher education affords the subject respect, dignity, and authority. This makes memorials every bit as much about values, status quo, and future as about remembrance." Beaver Stadium maintains the long-standing recognition of James Addams Beaver, one of Penn State's founding fathers, and that should not be lost, for his historic importance to the University is unquestionable. The stadium is also surrounded by memorials and traditions that remind us of a legacy of achievement and honor, as well as student and alumni pride that contributes to the national reputation of the institution. The removal of the Paterno statue and courtyard remains controversial as to what kind of statement it makes about values, honor, and dignity. The events that followed the Sandusky scandal will eventually

FIGURE 168 Sue Paterno at the installation of the Joe Paterno statue, November 2, 2001

reach their legal and institutional conclusions and, in time, historical perspective will be gained.

In the meantime, football and the events that surround a football game day remain cherished traditions but have also undergone much change over the life of the stadium and its predecessors. It is time to turn to the continuity and change that are encompassed in those experiences.

11

CONTINUITY AND CHANGE ON GAME DAY

Attending a game at Beaver Stadium is an unforgettable experience for both Penn State and visiting team fans. But there have been substantial changes over the years that affect how fans experience all the aspects of the game day.

Arriving, Tailgating, and Entering the Stadium

As one approaches Beaver Stadium today, it is hard not to be impressed by the massive structure that rises from parking lots where pastures once lay. It cannot be called beautiful, although it is monumental. Only the south end and the elevated three-story east side of the stadium have been transformed into an enclosed building. Otherwise, its structural elements are exposed to view, inviting fans to appreciate how the stadium has grown from a simple steel grandstand to the massive structure it is today. Even with its skeleton on view, it is still beloved as a place of social exchange and athletic excitement, embedded into the larger campus.

Since the days of its predecessors, there have been dramatic changes in simply getting to the game. In the 1950s most fans walked to the 27,000-seat New Beaver Field; today more than 35,000 cars surround the 107,000-seat Beaver Stadium. Football games, before television dictated starting times, were always a Saturday afternoon affair. They normally started at 1:30 p.m., and students with Saturday morning classes had to hurry to get there on time. Usually, a walk to the field to watch the game was preceded by lunch and followed by dinner on campus or downtown.

The move to Beaver Stadium introduced new circumstances. Almost 60 percent more spectators were in attendance than had filled New Beaver Field. While many came to a home game by car or bus, it was still quite a distance to find someplace nearby to eat. Local accounts tell us that *Centre Daily Times* editor Jerry Weinstein suggested, "just pack up some food and have a picnic." Thus was born tailgating at Penn State.

From simply having some sandwiches and a thermos of coffee from your car's trunk or station wagon tailgate, folding tables and chairs, with tablecloths, plates, napkins, cups, flags, banners, and canopies in blue and white have become the norm at Penn State tailgates. Despite the uniform color palette, the food comes in every variety imaginable—from chips and dip or cheeseburgers and chicken on the gas grill to entire meals cooked in the kitchens of massive RVs.

Now that kickoffs can be at noon, 3:30, or 8:00 p.m., tailgate tables are loaded with everything from breakfast casseroles and grilled stickies in the morning to barbequed dinner specialties in the evening. Many without tickets just come for the party and continue to tailgate throughout the game while watching it on TV via satellite dish or streaming Internet in the parking lots.

Tailgaters may hear Penn State music from members of University vocal ensembles strolling through the parking lots, singing the fight songs or the alma mater and happily accepting snacks and drinks from grateful fans. If you are in the right spot, you might also see the Blue Band marching to the stadium and generally having a great time warming up the crowd. One very special event is to watch the team walk down Curtin Road from their buses to the stadium entrance, surrounded by cheering fans.

When it comes to parking, aerial photos of the 1960s show virtually no paved parking lots, and little around the stadium; however, the surrounding fields were covered with orderly rows of cars and the occasional busload of guys, often from some distant Pennsylvania city. Tailgate tables and lawn games were rare in those days: fans gathered around individual cars were fewer in number and just partaking of food and drink from their trunk.

One of the primary changes to the tailgating scene around Beaver Stadium (and football stadiums everywhere) has been the arrival of the recreational vehicle (RV). While the motorhome originated in the 1950s, it wasn't until the 1970s that fans began to bring RVs to football games in increasing numbers.

With RVs now approaching 45 feet in length, separate lots have been created to accommodate these vehicles. Game-day RV spaces can also be found in several of Beaver Stadium's parking lots, and a special RV overnight lot off Orchard Road permits self-contained RVers to enter as early as Thursday after 6:00 p.m. before a home game and remain until Sunday.

With approximately 35,000 vehicles heading to the stadium, managing traffic has always been a major concern for a game day. The move from central campus changed the transportation patterns of coming to a game and, to an extent, road systems since then have been designed to accommodate football traffic. In fact, it could be argued that football has played a significant role in the gradual development of four-lane expressways to and through State College.

The original spectators for college football were college students, and with their energy and enthusiasm they are still the core audience for any game. Nevertheless, the number of seats allocated to students, their location in the stadium, the price of

FIGURE 171 Drumline leads the Blue Band to the stadium

FIGURE 172 Parking at Beaver Stadium in the mid-1960s

and proceeding to the seniors and graduate students on either the east or west sides. These were general admission seats for each section.

Like all modern entertainment venues, Beaver Stadium has come to rely on information technology to organize and operate its ticketing process. The devices that scan barcodes on tickets are part of a vast information technology system that coordinates a variety of services ranging from a student's paying for a hot dog at a concession stand with meal points to the complex systems that control videos and graphics shown on the scoreboards and electronic ribbons. And while tickets are now scanned, not torn, Intercollegiate Athletics still provides every buyer with a printed paper ticket and every vehicle with a printed paper parking pass. Counterfeit tickets and passes remain a concern.

Penn State uses a first-come, first-served system for the distribution of student tickets. Before the season, students have days assigned by class to enter their ticket requests online. As with class schedules and other essentials at Penn State, time is of the essence, and the 21,000 seats reserved for students quickly sell out. The game-day procedure enables students who line up first to have their IDs swiped to gain entry to the stadium. The system can cause delays, and it sometimes appears that the student sections are not full by kickoff and that students are still filing in through the first quarter. This often happens with a noon kickoff, when a number of students linger at tailgate parties close to game time.

For students, however, one celebrated tradition of attending a Penn State football game began when they started a 100-tent campground outside the student gates before the 2005 game versus sixth-ranked Ohio State. They dubbed it "Paternoville," and it has become a memorable experience that students check

student tickets, and the method by which they are distributed have undergone many changes, often with arguments over proposed modifications.

In the past, all nonstudent tickets were printed and torn at the stadium gates to allow one to enter as a spectator, while students merely had to show their student identification and matriculation cards to go in. Nonstudents were first charged for games in 1910; it was quite a bit longer before students paid to enter. In 1966 the first student season tickets were put on sale at $10.00. Student sections were arranged by class, beginning with freshmen in the end zone

off on their list of "things every Penn Stater should do before they graduate." Camping out is the way to be first to occupy front-row seats all along the student section. Of course, "occupying seats" is a relative term, since recent student tradition dictates that they stand throughout the game.

Gradually the process was formalized and a student group, the Paternoville Coordinating Committee, became responsible for organizing the campout. Now called "Nittanyville," the group has its own website, which explains the complex rules and provides camping information. Student groups often begin to stake out a place on Wednesday mornings before a game, and tents must disappear by 7:00 a.m. on game day. No alcohol, grilling, generators, or long extension cords are permitted, but the University does provide a power outlet to recharge electronics.

Over the years players and coaches have regularly visited the tent city, local merchants have provided food, and moms have baked cookies. It has even been featured on ESPN and in the *New York Times*. Camping out at Nittanyville has become a milestone experience for Penn State students.

In years past Beaver Stadium was completely open. One could wander in partway through a game without a ticket, stand along the fence that separated the walkway from the running track, and watch the end of the game. Although the stadium was not fenced, game-day access gradually became more regulated. But the events of 9/11 heralded a new era of security.

Places such as sports venues are considered potential targets for terrorism, and after the 1999–2001 expansion Beaver Stadium became completely enclosed with high barriers around its perimeter. Even delivery of supplies to the stadium for concession stands is under tight control. To enter the

FIGURE 174 Nittanyville campers

stadium at any time, visitors and workers alike must be on an approved list, use a designated gate, and show a photo ID. Spectators for the games are scanned with handheld metal detectors and checked carefully for what they bring into the stadium, and most types of bags are prohibited.

Inside the Stadium

While all fans anticipate the excitement of arriving in the stadium, the student sections are in many ways the most important part of the crowd. The students cheer the loudest. They start the wave, create the signs, dress and even paint themselves in blue and white to bring the spirit of the event to life, especially in a "white-out" game. In looking at photos of games from various eras, the increasing informality of the

FIGURE 175 Typical concession stand: front (LEFT) and interior (RIGHT)

spectators is striking. Where once students, as well as adults in the crowd, dressed formally, game fashion nowadays has become overwhelmingly casual and dominated by Penn State colors, emblems, and messages.

Another significant change is that spectators want more comforts and amenities for their day at the game. There are stands both outside the stadium and within where souvenirs and clothing may be purchased. And there are many more locations of concession stands than in the past, because people don't just go to a football game to see the action; they go to socialize, eat, and drink. At one time fans were satisfied with a hot dog and a Coke, with maybe a bag of peanuts or a candy bar for a snack, but the variety of food available has changed greatly over the years. The traditional hot dog has been augmented by more options, including burgers, chicken, bratwurst and sausage sandwiches, pizza, chips, popcorn, nachos, and hot pretzels. In addition, there is bottled water, Pepsi, Gatorade, hot chocolate, coffee, and, at

separate stands, ice cream and frozen lemonade. The logistics of providing this variety of food service can seem overwhelming. Ordering, inventory, distributing, and marketing all the products that go into the concessions depend on the nature of the items involved. Delivery and stocking of nonperishable items begins in the early summer, while perishable items are ordered on the Monday and delivered on the Wednesday before each home game. Even within the stadium, perishability dictates the method of distribution to the concession stands.

Concessions are under the control of Intercollegiate Athletics. Forty-eight stands of various sizes, with product variety depending on the capacity of the stand, are staffed by nonprofit organizations, about half of which are student groups, along with others such as church groups. A number of groups are from outside the State College area, some as far as an hour away.

The organizations that run the stands are supervised by about thirty monitors, who are part-time employees of Athletics. The organizations enjoy a 13 percent return for their efforts, and they get one additional percent of any sales by their volunteers who go out and sell in the stands.

It takes skill and experience to run a concession stand efficiently. Rushes occur before the game and at halftime; stands generally stop preparing food early in the third quarter. Weather, the time of the game, and the opponent can all affect sales. On hot days in the early fall, the sale of bottled water has reached as high as 70,000, compared to normal game-day sales of 15,000–20,000 (in all kinds of weather, water outsells soft drinks). On cold days, warm food and hot chocolate are the big sellers. Hot dogs are still the favorite food item, totaling 8,000–10,000 per game. Games that start at noon produce the greatest sales, with 8:00 p.m. contests yielding less, and 3:30 p.m. events the least. Close games keep the fans in the stadium and boost sales, while blowouts lead to an early departure and decreased sales. Of course, the concession stands are not the only sources of food and drink in the stadium. Spectators in the Mount Nittany Club and the suites have their own food service.

As the stadium has expanded, walking around the grandstands and access to the upper levels has become easier, and accommodations for the handicapped are now part of the stadium's seating in a way they never were before. Restrooms are more numerous and conveniently located. Information technology, which supports the wide variety of services from ticketing and entry control to media connections and communication throughout the stadium, has become a largely invisible backbone for managing a game.

A vital but relatively unnoticed part of the stadium environment is the presence of significant numbers of police officers, EMTs, and stadium security personnel. A command post in the press-box area can issue a call for help to meet any crisis or need. Whether it relates to the facility itself or a personal problem, a rapid response is possible, and with a venue as complex as Beaver Stadium, with over 100,000 patrons, there are frequent calls for assistance—whether it's an overflowing sink or a medical emergency. It has been said that there is no better place to have a heart attack than Beaver Stadium, and there are numerous accounts that attest to this fact. Well-trained first-aid people and doctors from Hershey Medical Center are present to react quickly to an emergency. Ambulances are on call, and it is a short distance to the Mount Nittany Medical Center's emergency room.

The Sound of the Game

Through most of the twentieth century, stadium noise came from the crowd, the band, or the announcements over the PA system. A coed cheerleading squad led fans in traditional cheers, but there were no dance numbers, the wave, or "We are . . . Penn State." The band, all-male with no majorettes or silks, was the only source of music. Of course, some traditions do continue—the Nittany Lion mascot was just as important in the 1940s and '50s as he is now.

A half-century ago, spectators at a game relied on the PA system to provide the details of the play of the game. A public address system was added at New Beaver Field in the 1920s. The Blue Band and cheerleaders go back to the turn of the century, and crowd reaction to the game was fundamental to what was taking place.

Since the move to Beaver Stadium, the Blue Band's role in the game has grown. In 1960 the band

FIGURE 176 Blue Band on New Beaver Field, early 1940s

of the south end-zone tunnel at its frenetic 180-beats-per-minute quickstep to form a block of long lines stretching across the field. As they play their well-known fanfare and break into "Downfield," the band steps off and the drum major begins his race to "the flip." As the band continues, the drum major reverses field, races toward the student section for his second flip and rendezvous with the Nittany Lion, who presents the drum major with his baton.

Meanwhile, the band moves into formation to play the "Star-Spangled Banner." After the national anthem and an abbreviated version of the visiting team's fight song, the band, playing "New Fight on State" then moves into a block PSU and plays the Penn State alma mater. Then, playing "Nittany Lion" (often mistakenly called "Hail to the Lion"), the band forms the block-letter "LIONS" while marching up the field. This "Floating Lion" maneuver is completed as the band seamlessly reverses field at the north end zone and continues the formation with "Fight on State" until it reaches the south end zone. The band then forms the aisle through which the team runs onto the field from the tunnel. No true Penn Stater can fail to be stirred by this program.

Collaborating with the Blue Band, the cheerleading squad, now coed and considerably larger than it was in the 1950s and '60s, also plays a significant role in the pregame. The "Mike Man" warms up the crowd as the players gather in the tunnel. When the team and coaches run onto the field, they are led by the cheerleaders, with members of the squad bearing the nine large blue and white flags, each with a letter, which together spell out "P-E-N-N-S-T-A-T-E."

Cheerleaders now spend most of the game in front of the student section, working with the Lion, leading the crowd to cheer and occasionally do the wave. For part of the game, they exchange places with

had 108 members; today's coed band also includes the Blue Band Silks, the Touch of Blue majorettes, the Blue Sapphire feature twirler, and the drum major, and totals about 275 members. The more the stadium expanded, the more the band had to grow so that they could be better seen and heard.

The Blue Band has also grown in the sophistication of its halftime shows over fifty-plus years. Shows are generally "themed" and include music and field formations to match the chosen theme. Every new show requires both novel choreography and good band arrangements for the music. Over the years, the formations and the transitional moves between them, which take several days to plan, even with the use of computer software, have become more complex.

Nevertheless, it is the pregame show that may well be the most anticipated part of the band's performance. Director Ned C. Diehl developed the routines in 1965. Following the stately entrance of the drum major and percussion section, the band explodes out

the Lionettes dance team, who are usually stationed in the north end zone, also leading cheers. Both the Lionettes and the Lion, with some of the cheerleaders, do special dance routines during the course of the game. The cheers one might have heard in the 1960s, such as "Short-yell State" or "The Nittany Lion," were gradually forgotten, and outside of the crowd's cheering for good plays and scores, there was a noticeable lack of coordinated or rhythmic cheering.

The famous "We are . . . Penn State" cheer started in the mid-1970s. The cheerleaders were impressed by the great power of rhythmic call-and-response cheers they heard at Ohio State, USC, and Kentucky and developed Penn State's signal cheer. After several years, the crowd finally adopted it in 1981, along with the "Thank you. . . . You're welcome" conclusion. Today, you might hear it anywhere one Penn Stater encounters another.

After the game ends, the football team, win or lose, moves to the southeast corner and joins with the band and the student body to sing the alma mater. Then as the team leaves through the tunnel, the band returns to the field to do its postgame show.

Today's Beaver Stadium games are loud. According to a Penn State acoustics expert, recorded music and crowd noise reaches from 100 to over 120 decibels at various times during a game, especially in front of the student section. In addition to the sound of the powerful audio system, the massive video scoreboards and electronic ribbons (or banners) that are attached to the faces of the north and south decks pulsate with animated announcements and advertisements for corporate sponsors. Television timeouts have extended the crowd's time in the stadium, and the intensity of sound and video now barely leaves a moment when fans are alone with their thoughts. Television monitors even enable those standing in

line at a concession stand or in a restroom to stay current with the action.

All the "visual excitement" presents an experience not unlike what people recognize from their personal electronic devices and social media. Band directors call the "over-the-top-entertainment" expectations of the crowd a major challenge. Even with microphones, the band and cheerleaders cannot be as loud or visually stimulating as all the new technology.

The intensive use of media draws the crowd's attention to advertising, video, and audio whenever the teams are not actually engaged in playing the game. This orchestrated game more reflects society today. It is just one of the changes that many attribute to Guido D'Elia, who served as "Director of Branding and Communications" for football from 2004 to 2012. He had been involved with televising and promoting Penn State football since 1979, including producing the Emmy-winning *Penn State Football Story*.

FIGURE 177 Blue Band forms aisle to welcome the team onto the field for the Michigan game, November 21, 2015

When D'Elia joined Penn State full time in 2004, one of his responsibilities was "live game-day production and orchestration of the atmosphere inside and out of the stadium." He suggested changes to the look and promotional signage at the stadium. His pre-game videos dramatically portrayed the team's spirit and success. His initiation of white-outs and the use of high-volume music like Zombie Nation's techno song "Kernkraft 400" and stadium anthems like Neil Diamond's "Sweet Caroline" have boosted crowd participation and enthusiasm. All of these innovations have led commentators at ESPN and elsewhere to refer to Penn State home games as the "greatest show in college football," although not without some grumbling from those who prefer most of the excitement to be on the field.

The Media Cover the Game

Penn State's first football game in 1887 was covered at length in the *Free Lance*, the first continuing student newspaper. In the more than 125 years since then, there hasn't been a Penn State football game that was not written about or, since the introduction of radio, broadcast to the public. Looking at the game coverage on television, radio, newspapers, magazines, and on every imaginable form of Internet and social media, it is hard to imagine almost any aspect of the game that is not the subject of some form of commentary.

Nevertheless, 200 to 300 members of the press who receive game credentials from the University's sports information officials can still be found in the press box and on the sidelines reporting on and photographing the game. Depending on the strength of Penn State and their opponent on any given day, the number of press covering the game can vary considerably. The days

when the few reporters could sit down with a coach or player to informally talk about the game seem long gone. Just as in politics and other areas of media coverage, the message is now carefully managed.

Beaver Stadium today provides a variety of resources and amenities for reporters in the press box and in the media rooms outside the locker rooms. It was not always so. New Beaver Field's first wooden press box was erected in 1924, and it was spartan by all accounts, with barely enough space to sit and pound a typewriter or take notes. As the field transitioned from wood to steel in the mid-1930s, the press box was extended to provide more room for reporters and for college officials and guests.

When the seating capacity of New Beaver Field was doubled in 1949, the press box was completely redesigned as a four-level structure that was a state-of-the-art facility for its time. In the 1959–60 move to the east end of campus, the press box was also taken down and moved along with the grandstands to the new location. At its new height, however, Penn State added an elevator to make it easier for reporters and broadcasters to reach the box.

The press box was expanded again in 1969, although most of the new space, which doubled the length of the box at its upper three levels, was to provide more space for VIP seating. When the press box was lifted 12.6 feet in 1977–78, along with the rest of the stadium, additional renovations were made. In 1980, during another incremental expansion of the stadium, the press box was extended again to provide more space for broadcasters, the print media, and VIPs, and two elevators were added.

When the 2001 addition was completed, adding the south decks and east-side luxury suites, no changes were made to the press box. Television play-by-play announcer Keith Jackson, broadcasting the first game in the newly expanded stadium, is said to have commented that Penn State spent $93 million on its addition, but didn't put 93 cents into improving the press box. Its challenges for power, data, and broadcasting needs have long been recognized, and in 2015 the Board of Trustees approved a long-overdue upgrade of the press box at a cost of $2 million. This included replacing the south elevator, changing out windows, carpeting, lighting, and other finishes, as well as improving restrooms, emergency lighting, and interior painting.

Few out-of-town reporters made it to State College in the early days of Penn State football, and much of the coverage for other papers came through Penn State students acting as stringers. But as Penn State launched its fundraising campaign in the early '20s, it also began a public information office, and with Hugo Bezdek's success a full-time sports publicity staffer was hired in 1922.

The sports information staff now creates a large annual media guide and game programs, generates press releases, runs press conferences, handles player relations, assists with television production, and handles media credentials. The sports information director is now associate athletic director for strategic communications, with a staff of a dozen or more to cover all Penn State sports. In addition, there are related units for creative services and brand management, marketing, and media and video production.

While many well-known sportswriters covered Penn State football, perhaps the most notable pair of writers to have witnessed virtually every game wrote the *Alumni Association Football Letter*. Ridge Riley, the eventual executive director of the Penn State Alumni Association, began a weekly letter for distant alumni in 1938 and continued it up to his death in 1976. Since that time John Black, a former *Collegian* sports editor and

FIGURE 178 Beaver Stadium press box at the homecoming game vs. California, 1966

long-time Alumni Association staff member, has continued the *Football Letter* for more than four decades.

Radio broadcasting began in the 1920s, with a commercial radio network created in 1938. Football's popularity with the public made broadcasting Penn State games a major form of outreach for the college. Penn State played its first game on national television in 1954. By 1975 Penn State was providing delayed broadcasts of its games over eleven channels in four states. Now, through the Big Ten network and other network contracts, televised games have become a significant source of revenue for athletics and every Penn State game can be seen on television.

But for real fans of Penn State football, nothing can replace our own radio network broadcasters. There

FIGURE 179 Ridge Riley (LEFT, IN GLASSES) and Lou Bell (SECOND LEFT), head of public information for Penn State, in the press box at New Beaver Field, 1950s

FIGURE 180 Bob Prince (RIGHT) in the booth at New Beaver Field, with Mickey Bergstein (SECOND RIGHT), general manager of WMAJ radio and later a faculty member in the Smeal College of Business, serving as his "color man"

have been many over the years, including the well-known Bob Prince, famous as "the gunner," the voice of the Pittsburgh Pirates. In 1970 Fran Fisher began doing play-by-play, retiring in 1983, but coming back in 1994. He was teamed initially with Mickey Bergstein and later with George Paterno, the coach's brother, doing color, and the pair became a Penn State institution. Fisher retired again in 1998 and Steve Jones has handled the microphone, along with Penn State Hall of Famer Jack Ham providing color since then.

Preparing the Stadium for Game Day

There have been periodic structural changes to Beaver Stadium over the years, but ongoing maintenance is always needed for any structure. While some of this is of a housekeeping nature, there are other needs that, although not significant enough to be considered

expansions, affect the appearance or functionality of the venue.

Preparing the stadium for the fall schedule of games involves the reactivation of several systems following the winter shutdown and is done by the Office of Physical Plant (OPP). Except for heated areas, which are part of the 1999–2001 expansion and used all year long, all of the pipes in the remaining portions of the stadium are drained after the last game in the fall. Then, starting in the late spring, the process is reversed; the water pressure is restored and the entire system is checked. In addition, through the summer months, all seating areas, concourses, concession stands, and restrooms are cleaned and resupplied in readiness for the first game. The areas immediately outside of the stadium are also cleaned and trimmed, including the mowing of the near-stadium parking fields and the placement of the parking number markers. These are done by the crew that prepares the playing field, but the other parking areas are maintained by OPP.

Routine maintenance of the structure is also necessary. This might involve the replacement or tightening of loose bolts in the decking, occasional sinkhole and foundation repairs, and the overall inspection and repair of the supporting structural members. Periodic inspections are done to monitor corrosion of the steel decking. There is also interior painting and recarpeting and, over time, the wall murals and posters along the concourses or within the recruitment area and Letterman's Club are changed to reflect both the historic and recent achievements of the football team and its players.

A host of minor construction jobs are addressed each year in the relatively short window between seasons. Over the years they have affected the nature of the facility, such as replacing the wooden bench seating with aluminum, upgrading electrical power connections, and the construction of a stone wall at the lower edge of the stands.

Major projects are also done by outside contractors under the supervision of OPP during the off-season. These would include sandblasting and painting the underside of the steel stands, installing permanent lighting (which was done in the 1980s), and checking or retensioning the tendons on the corbels that support the entry ramps of the north deck.

Another of the most recognized features of Beaver Stadium is the surface on which the game is played. While many university and professional football venues have installed artificial turf, Penn State has continued to play on its renowned natural turf, which football players, coaches, and commentators agree is one of football's finest playing surfaces. This is done at significant expense and with many challenges, but it should come as no surprise; Penn State's College

of Agricultural Sciences features one of the nation's premier turfgrass programs and graduates manage the grounds at numerous professional sports stadiums and top golf courses.

Many of the turfgrass students receive hands-on experience working part-time with Beaver Stadium's experienced grounds crew. Maintaining the field requires a careful program of mowing, watering, fertilizing, cultivation, and pest management. Before home games, the crew spends the better part of a day painting the lines, numbers, and symbols on the field.

Penn State's field is planted in a mix of Kentucky Bluegrass varieties, the best athletic field grass for our climate. Most fans will have noticed that the use of Beaver Stadium's field is tightly controlled. There have been no concerts or other events staged on the field, football practices in the stadium are strictly limited, and the Blue Band marches on the field only on relatively dry game days. Many in the past attributed this to a Coach Paterno obsession for keeping the field pristine for football.

Actually, the turfgrass specialists have recommended the stringent limitations on the use of the field. These practices compensate for the poor design

FIGURE 181 Painting the field in preparation for a game

of the field's subsurface when it was first laid down in
1959–60, in which overly compacted soil has caused
severe drainage problems. The original construction consisted of a six-inch-thick layer of gravel laid
directly on the subsoil that contained perforated
drain pipes spaced six feet apart, which was standard
practice. On top of the gravel, however, the builders
placed an eighteen- to twenty-inch layer of silt loam
soil—essentially topsoil—on which was rolled out a
layer of Kentucky Bluegrass sod.

The depth of the silt loam soil was the problem.
Water could not move through it to the gravel, leaving the soil completely saturated and unable to drain
during rainstorms. In conditions like this the sod
layer could not bond to the soil and withstand the
stress of heavy football players making sharp turns,
starts, and stops.

A number of fixes and procedures were developed
to preserve the turf. A ten-inch crown was created
down the middle of the field (noticeable to those on
ground level) to promote runoff to the sideline where

catch basins collected and transported the water to
subsurface drain tiles. The field is carefully mown
and watered to maintain the proper balance of play-
ability and drainage, and it is routinely covered with
a tarpaulin whenever there are heavy or long-lasting
rainstorms to protect the underlying sod structure. As
long as these conditions remained, it seemed unlikely
that the field would be used for concerts, outdoor ice
hockey games, or other events in the future, because
they would likely require replacing the turf.

Perhaps the most infamous game from a turf
standpoint was the 1993 Rutgers game. The turf
had been replaced with new sod in July, but with a
hot, dry summer and a possible root disease, it was
already breaking up and becoming dislodged by the
beginning of the season. The game against Rutgers
took place after several days of rain. Although the
field was covered until just before kickoff, it rained
heavily during the game, and large chunks of turf
were being dislodged with every play. With an away
game and a bye week, there was time to bring in new

sod and replace the turf. This lasted to the end of the season, but it was clear that short-term fixes were not working.

The solution was to drill twenty-inch-long holes into the field, which were then filled with sand, enabling water to reach the gravel and drainpipes. Fixing the entire field, however, required more than 100,000 of these sand columns, which took several years to complete. The alternative, to excavate and rebuild the entire field using a base of sand instead of silt loam soil, was considered too expensive because of the impossibility of bringing large earth-moving equipment into the bowl.

Late fall games can bring another kind of "white-out," and you'll see the grounds crew sweeping snow off the yard lines during timeouts. Snowfalls before games, like the seven inches of wet snow that fell the night before the 2009 homecoming game, may require fans to use parking lots around town and shuttle buses for transportation, as the many grass lots become unusable.

In the long years of Penn State football history, two Beaver Stadium snow games are legendary. One was the November 18, 1995, game against Michigan. The storm occurred midweek, dumping 18–20 inches of snow, and winterlike temperatures persisted through the rest of the week. Instead of canceling the event, Intercollegiate Athletics decided to press on with the game.

The challenge was to have the field ready and the stadium prepared to safely host nearly 100,000 people. As for the playing field, since Beaver Stadium was a closed bowl, large snowplows and trucks could not be brought in to remove the snow. Therefore, more compact front-end loaders were used to heap huge piles of snow along the sidelines and beyond the end zones. Hundreds of volunteers, including inmates

from local correctional centers, were used to clear the stands.

Remarkably, an estimated crowd of 80,000 turned out for the game. Like the unexpected size of the snowfall, the Nittany Lions also provided a surprise at the end of the game. With a little over two minutes left, holder Joe Nastasi overcame the freezing conditions and scored a two-yard touchdown off a fake field goal to seal a Penn State win by a score of 27–17.

The other noteworthy snow game took place forty-two years earlier, when Penn State almost forfeited a home game because it couldn't get to the stadium. An unanticipated early winter snowstorm blanketed central Pennsylvania on Friday, November 7, 1953, depositing seventeen inches before it ceased. Rip Engle had decided to take the team out of town the night before the game to sharpen their focus. They became snowed-in at a hunting lodge in a rural part of Clinton County, about twenty-five miles from State College.

FIGURE 184 Front-end loaders move snow to the sidelines before the Michigan game while volunteers and others clear snow from the stands, November 18, 1995

FIGURE 185 Clearing the snow before the
Fordham game, November 7, 1953; note
heavy road grader plowing the snow

The Penn State team arrived on campus just in time for the game's 1:30 p.m. start. Ironically, the mountain lodge where the team had stayed was named "Camp Hate-to-Leave-It." But it all worked out for the Lions, and Penn State won by a score of 28–21 before a hardy crowd of 13,897.

While maintenance and construction projects are handled by University employees, many of the operations that make a football game possible are handled by volunteers. At the lot entrances, parking staff control entry, collect fees, and direct cars to reserved or open spaces. The parking attendants receive a small stipend, a food allowance, and a bleacher seat in the north end of the stadium for their efforts. The parking staff remains until two hours after the end of the game to direct and aid motorists, which, for an 8 p.m. kickoff, might keep them in the parking lots until well after midnight.

Volunteer ushers also help patrons locate their seats and, with part-time security staff, they keep a watch on the crowd to spot emergencies or other problems. They are backed up by police and EMTs.

More than a thousand volunteers work the concession stands. These volunteers are expected to arrive four hours before game time and must be sixteen or older. During slack times at the concession stands, volunteers from the stand can serve as hawkers in the seating areas. They augment a group of "professional hawkers" who operate under a separate contract. Volunteer hawkers can offer any of the items sold at their respective stands, whereas the professionals only sell drinks or programs for fans who did not purchase one on their way into the stadium. Buying a program is much less common than in earlier times. Thanks to the media, the electronic scoreboards, and cellphones, fans can learn all they might want to know about a game before, during, or after the contest.

On Saturday morning, the visiting Fordham team was in town, but the Nittany Lions were stuck in the mountains. Once the bus to transport the team got close to the lodge, the team hiked out through the heavy snow. The story goes that Rip Engle led the way, followed in single file by the other coaches, second- and third-team players, and the linemen to break a path of minimum resistance for the backs. They eventually reached the bus, and the state police escorted the team back to town.

On campus, large road graders were used to plow the snow to the edges of the field, while trucks carried it away out the open south end of New Beaver Field. When the temperature moderated that afternoon, the residual snow on the field melted, leaving the playing surface in good shape by game time. Concurrently, hundreds of students and other volunteers cleared sections of the stands, heaping the snow in large strips at the ends of the cleared portions.

Routine after-the-game cleaning is done for compensation by a host of student groups. Some of these groups are athletic teams of minor or club sports that use this opportunity to accrue financial support for activities not covered in their budgets from the Athletics Department.

For more than sixty years, Beaver Stadium has hosted Penn State football games and, as it expanded, the stadium has come to be an ever more significant part of Nittany Lion fans' affection and enthusiasm for attending a home football game on campus. From tailgating and the opening Blue Band show to packing up to head home, traditions and heritage are interwoven throughout game day in both the look and feel of the stadium as well as the customs and behaviors of the crowd while the game takes place. As the success of the football program, and its importance to the University, has grown over the decades, the size and character of the stadium have evolved to make the home-game experience the "greatest show in college football."

12

BEAVER STADIUM'S IMPORTANCE AND FUTURE

The story of Penn State's football venues and the memories they hold and embody is anything but one-dimensional. We've looked at the story from both the standpoint of the physical structures where fans assembled to watch football games and how they enjoyed those games over the decades. But there is also the larger context of the University, the campus, and role of football in that history to consider.

The Importance of Football for the University

In a state that was well populated with private colleges and a system of state teacher-training institutions, Penn State had stiff competition for both students and state appropriations in the nineteenth and early twentieth centuries. For many Pennsylvanians, it was

the "cow college," or at the very least it was known as that small college up in the mountains, "equally inaccessible from all parts of the state." The nascent Pennsylvania State College used sports, and football in particular, to help build public awareness of its potential as well as recognition of its status as Pennsylvania's land-grant college.

As Penn State emerged from its "dark ages" in the 1880s, it slowly built an academic reputation in engineering, mining, agriculture, and the applied sciences. True to its land-grant mission, it did not exclude the liberal arts and eventually gained recognition for those programs as well as education, home economics, and the practical arts.

The quality of its instruction and its research programs also became better known as Penn State spread geographically across the state. Beginning with the establishment of the network of county agricultural agents in 1913, Penn State's extension programs soon spread with the creation of specialized centers for instruction in mine safety, teacher training, foreman training, and engineering technical schools in several urban industrial centers. During the 1930s Penn State opened extension credit centers in communities that were particularly hard hit by the Great Depression; these centers became the nucleus of today's Commonwealth Campuses.

The public's recognition of Penn State's outstanding performances in athletics complemented these efforts. In that era, such accomplishments demonstrated that a school developed capable students, where athletic prowess was evidence of strong character and the qualities that would lead to success in life's pursuits after college. The athlete was idealized by the public; he was an object of admiration and emulation. In a state, and a nation, with literally hundreds of colleges and universities, sports gave Penn State opportunities to compete with elite schools at a time when the nation's finest academic institutions did not find it contradictory to aspire to excellence on the playing fields as well as in the classroom.

In 1958 Penn State accepted membership in the prestigious Association of American Universities, which marked its recognition as one of the nation's major research universities. Its success as an academic institution was now unquestioned. However, outstanding athletic programs remained an important ambition for Penn State. They cemented the loyalty of its growing alumni base, attracted increasing numbers of students, and contributed to the pride and support of Pennsylvanians and the state government.

By the early 1970s the extraordinary success of Penn State's football program under Coach Joe Paterno became the primary force driving the expansion of Beaver Stadium. Over three decades, winning records, repeated bowl game appearances, two national championships, and a successful integration into the Big Ten athletic conference all contributed to increased demand for game and season tickets. At the same time, Penn State proudly endorsed Coach Paterno's philosophy of "success with honor" to recruit and nurture student-athletes who would achieve the highest graduation rates possible and have successful life experiences beyond whatever glory they might achieve on a gridiron. This also enhanced the reputation of Penn State athletics and the pride that students, alumni, and fans felt for the program.

Penn State's stature in intercollegiate football was more than mythology to Penn Staters, and those feelings underlay all of the celebration and pageantry that became part of the Beaver Stadium football experience. Media also played a vital role in promoting football at Penn State. Newspapers and especially radio broadcasting had allowed fans to vicariously share the excitement of the game, hearing the sounds of the band and the crowd, as their imaginations carried them to New Beaver Field or Beaver Stadium. Television intensified that experience, and new technology enabled fans to watch any Penn State game via the Internet from virtually anywhere in the world. This has built enormous fan identification with Penn State across the nation and beyond.

At the same time, football also played an important role in the finances of the University. Before World War II, there can be little doubt that Penn State existed on a tight budget. Regular state appropriations did not begin until 1887, and even after that the institution's administrators had to regularly make the case that it was a public institution deserving of state funding. Without alumni of means to make substantial donations, it received only two major gifts in the nineteenth and early twentieth centuries—the buildings that bear the names of their donors, trustees Andrew Carnegie and Charles Schwab.

The recycling of grandstands is a perfect example of how Penn State dealt with the financial constraints under which it operated. The grandstand at Old Beaver Field was taken down, moved, and reassembled at the northwest corner of campus in 1908 and, beginning in 1934, those wooden stands were gradually replaced by steel grandstands. Twenty-five years later, New Beaver Field's steel stands were taken apart, moved, and reassembled at the current location. This move mated the existing grandstand with 16,000 new seats already constructed on site. Together, the new 44,000-seat grandstand and field was rechristened Beaver Stadium.

All of the existing steel stands erected between 1934 and 1972 were built using the patented Lambert Grandstand method. This allowed preformed sections of steel decking to be bolted together to make any length or height of grandstand needed, thus providing maximum flexibility to meet the growing needs for expansion, as well as possible relocation, in an economical manner. Only with the major expansions from 1977 to 2001, which enlarged seating capacity to more than 106,000, were more complex grandstands and supporting amenities created and paid for with football revenue.

The college's 1930 master plan, along with successive plans through the early 1940s, proposed a permanent stadium on the western edge of campus. However, with Depression-era budget cuts and wartime and postwar stringency, it was never built. The grandstands, both wooden and steel, were recycled and moved when necessary. Penn State, a practical, utilitarian institution, had to operate in the most fiscally prudent manner possible.

In its earliest years, Penn State tried to interest alumni and friends in donating funds to support the institution. The Farmers' High School depended on gifts of land, farm equipment, plants, and animals, as well as cash, to help it get off the ground. By the turn of the twentieth century, trustees were the most frequent donors. The first organized fundraising campaign was launched in 1922 by President John Martin Thomas. His Emergency Building Fund drive sought out donations for new gymnasiums and housing for athletes among other buildings for students. Even then, Thomas appreciated the value of football as a way of attracting the interest of donors.

But it is only in the modern era that football became so closely tied to fundraising. Penn State instituted a formal fundraising organization through the Alumni Association in 1952 with the Penn State Alumni Fund, but this was only a modest effort compared to what would come later. By the 1980s a large and well-organized development operation had been created to oversee the first modern fundraising drive, the Campaign for Penn State (1984–90), which raised $352 million.

Athletic fundraising also played its role in the new development efforts. The Levi Lamb Fund was created in 1953 as a focus for gifts to support athletics, and in 1961 the Nittany Lion Club became the overall fund to coordinate gifts to support teams and scholarships and to serve as a mechanism for handling football-ticket distribution and recognition for donors. Since the Campaign for Penn State, general campaigns like the $1.4-billion Grand Destiny Campaign and the $2.1-billion For the Future Campaign, completed in 2003 and 2014, respectively, as well as targeted efforts for specific projects like the Beaver Stadium expansion have become the norm.

In these efforts, football is not just a recipient of gifts. Ever since the expansion of the stadium's press box, increasing numbers of VIP seats have been created to provide a means to entertain and cultivate potential donors to the University. With the addition of the luxury boxes in the 2001 expansion, in particular the President's Suite, thousands of donors and friends may be entertained over the course of a season.

The typical home-game Saturday begins with the "Huddle with the Faculty" programs to showcase academic achievements. Invited guests proceed to the President's Tailgate and on to the VIP seating at the game, and then to postgame receptions and parties hosted by college deans and department heads. Comparatively few of these guests will likely be donors to the athletics program, but this football game-day experience has become a critical part of the massive fundraising campaigns that are vital to advancing the work of the University in a multitude of ways.

All in all, football has come to play an integral role in the life of the University in terms of attracting students to enroll, supporting one of the largest college athletic programs in the country, being the stage for both informing and entertaining legislators and government officials, and cultivating donor relationships with alumni and friends.

Beaver Stadium: Not Just for Football

Of course, when one thinks of Beaver Stadium, the first thing that comes to mind is the football game that is played there. However, its presence richly enhances the life of the University and the community in ways that range over a spectrum of functions.

In the area of athletics, as was true for the earlier Beaver fields, the stadium served as home for track and field meets until the installation of permanent bleachers in the open south end in 1976. Since that time, the Blue-White game at the conclusion of spring football practice is the only public athletic use of the stadium besides the fall football games. Once just a scrimmage, it is now a four-quarter game that provides a chance for students and fans to experience a free game. It has also grown into a time for campus development functions, special events, and tailgating for as many as 50,000 fans who come to preview the developing Nittany Lion team.

Before the walled exterior that was constructed as part of the 2001 expansion, the open stadium also served as a gathering point for several of the summer sports programs. Parents would assemble under the shelter of the stands to drop off and pick up their kids, and the home locker room was the point of assembly for the participants.

Both New Beaver Field and Beaver Stadium have hosted the spring commencement ceremonies. The first graduation program held at New Beaver Field took place in 1939, and the spring 1960 graduation ceremonies were held in the partially completed Beaver Stadium. In the 1960s and after, thousands of graduating students sat around the north end of the stadium, while family and friends lined the east and west grandstands. Crowds of 30,000 or more were not uncommon in those years of booming enrollments. The last outdoor ceremony at Beaver Stadium took place in spring 1984, after which graduations moved indoors for smaller individual college and graduate school observances.

While football tailgating is the focus of the fall, other events have traditionally been held in the parking lots around the stadium. In the 1960s and '70s Spring Week featured both carnival rides and the highly anticipated, and generally hilarious, fraternity and sorority Spring Week skits, held in tents along a midway in the latter portion of spring term. In the 1970s the "Movin' On" celebration, at the end of spring classes and tests, evolved from a HUB lawn concert into a daylong event with carnival rides and continuous music in the adjacent intramural fields that surround the stadium.

While the paved parking lots around the stadium serve as commuter parking for students and staff during the week, they also accommodate year-round sporting and arts events. Perhaps the best known is the annual Central PA 4thFest, which began in 1991. Its forty-five-minute fireworks show uses computers to choreograph fireworks to music, firing over ten thousand shells and relying on the services of six hundred community volunteers. VIPs and major contributors to the event are invited to observe the festivities from the comfortable settings of the south and east portions of the stadium.

Beaver Stadium has been a closed facility since 2001, but prior to that both New Beaver Field's and Beaver Stadium's grandstands had long sheltered stadium groundskeeping and maintenance equipment and Office of Physical Plant equipment and supplies. In addition, vehicles used to conduct University business were also parked under the New Beaver Field stands.

Facilities at the stadium were also used in support of several academic programs over the years. Besides the Agronomy Department's turfgrass science studies associated with the stadium's playing field, the Civil Engineering Department conducted its undergraduate summer surveying program at the stadium for several years after the program moved onto campus from the Civil Engineering Camp at Stone Valley. The home-team locker room was used to store the surveying equipment and served as the base of operations

for the field activities that were conducted in the open spaces that surrounded the stadium in the 1960s.

The stadium also played a key role for the Department of Astronomy and Astrophysics in its joint venture with several other universities in the development of the Hobby-Eberly telescope. The original elevator shaft, which was vacated when the two new elevators were installed in 1980, was used to test the lenses for the telescope. This is the fourth-largest telescope in the world, and it is housed at the McDonald Observatory in the West Texas mountains.

Beaver Stadium was once the subject of a study by students from the Aerospace Engineering Department. Prior to the lift of 1978, the stadium had an opening at the northwest corner to accommodate some of the track events. There was a question as to whether this opening created wind currents on the field that would hamper the passing game. To study the problem, students built a wind-tunnel model of the stadium and studied the air flow on the field from wind currents that originated through the opening in the stands.

Today, the most significant event taking place under the grandstand is the Trash to Treasure spring sale. Penn State students moving out of dorms and apartments donate unwanted materials to the sale before they leave campus. An average of 68 tons of clothing, household items, appliances, bedding, furniture, electronics, and nonperishable foods are sorted and organized by five hundred volunteers. The sale netted about $50,000 in 2015, and since its initiation in 2002, it has donated over $660,000 to United Way partner agencies while reducing the amount of materials sent to landfills.

Since the enclosure of the south and east sides, the Mount Nittany Club space and the lounge spaces outside the suites area have been available to be rented for a variety of private events, such as wedding receptions, high school proms, and birthday celebrations. In some years, as many as 150 events have been held there.

Probably the best-known events, however, are the weekly fall luncheons of the State College Quarterback Club. The club was founded in the early 1950s by local businessmen with an interest in Penn State football. The Wednesday meetings, then held at the Hotel State College, featured Coach Rip Engle and often an assistant coach and a player. Engle showed game film, explained the action, and answered questions—off the record, as media members were not invited. Mickey Bergstein served as host for many years as the program became an integral part of the life of Penn State football. The club has grown to more than 500 members, but the luncheon programs, hosted by Penn State sports announcer Steve Jones, still feature remarks from coaches and players. The club also sponsors the annual football banquet at the Penn Stater to honor the seniors and maintains an endowment fund to support academic enhancement activities for players, including computers, academic counseling, and other needs.

For more than twenty years, arguments about holding more events in Beaver Stadium have accompanied discussions of additions and renovations. In the past, outdoor concerts were the most frequently proposed events, but since 2008, outdoor NHL ice hockey games have been occasionally held in football and baseball stadiums. In 2014 a crowd of over 105,000 attended a Toronto Maple Leafs–Detroit Red Wings game in Michigan Stadium. As a result, there have been frequent calls for an outdoor "classic" to be played at Beaver Stadium between the Pittsburgh Penguins and Philadelphia Flyers. Such uses have not been given serious consideration because of the long history of drainage problems that has required a stringent program to protect the turf. In fact, when the

superintendent of construction for the 1978 lift was interviewed, he said that they were forbidden from moving any equipment over the playing field and were even asked not to walk on the field. It seemed unlikely that the field would be used for other events, because they would likely cause major problems with the turf, not to mention the possible seasonal issues of reopening winterized facilities, concession stands, and other spaces for use. However, a concert has been scheduled for summer 2017.

Future of Beaver Stadium

Throughout the history of New Beaver Field and Beaver Stadium, there has been an ongoing array of studies on how to enhance the facility or increase its seating capacity. Some of the changes that were considered were never implemented; however, as outlined in the earlier chapters, many were incorporated in what is today's Beaver Stadium. The scope of these changes ranged from minor incremental adjustments to ones that changed the very nature of the venue. The most recent modest change involved renovations to the press box prior to the 2015 season. These were seen as transitional in nature, as it was anticipated that major changes on the west side would eventually occur. Also, in the spring of 2015, Penn State commissioned the Kansas City, Missouri–based architectural firm Populous to recommend improvements and funding plans for all of its eighteen athletic facilities, including options to "renovate or replace" Beaver Stadium. Renovation would add one more chapter to the story tracing the evolution of one of the nation's largest and most venerable college football venues.

When renovation is discussed, the west side of the structure is usually the focus. It is the one quadrant of the stadium that has not had a major change since 1978, and it is where potential plans for stadium renovation have been directed since the last expansion. Several conceptual options were presented by Populous in 2009 to replace the existing press box with a multi-level structure that would somewhat mirror the east-side suites. These conceptual plans are not currently under consideration, but they give a sense of what might be. The topmost level would house broadcasters and reporters and a limited number of suites, the next-lower level would be exclusively dedicated to suites, and below that would be a club space behind an outdoor club seating deck projecting toward the playing field. A number of small loge boxes to the rear of the club seats would also have access to the club space. All options included additional space for handicapped seating. One of these plans involved a slight increase in seating capacity, whereas the other options reduced seating.

In addition to enlargements and improvements to the west side, further stadium enhancements were also to be considered in each of the 2009 options: a remodeling of the field-level concourse around the entire stadium, including the replacement of antiquated ground-level restrooms, concession stands, and the visiting team locker room; adding elevators at key places to improve access to upper-level seating; improvements for handicapped seating at several locations throughout the stadium; and some visual enhancements. It is possible that one of these options would have been implemented over the following years if the Sandusky scandal hadn't occurred in 2011, but, as events unfolded, everything was placed on hold.

In University master plans over the last forty years, a major emphasis for the future of the campus has focused on renovating and upgrading existing spaces that have reached or surpassed their normal useful life. Such changes require long-term planning as an investment for the future. By some standards, parts of Beaver Stadium are little changed from the move in 1960, and from a physical standpoint the steel grandstands dating from mid-1930s New Beaver Field are still part of Beaver Stadium today.

Accessibility and comfort issues in the stadium are frequently discussed. Modern stadium aisles have railings down the middle for safety, and armchair seating is expected in any new stadium. Newer stadiums often use multiple escalators to move crowds to upper levels. Such retrofits in Beaver Stadium would, of course, require major changes. Modern professional stadiums built in recent years have provided the full range of amenities that Beaver Stadium lacks.

Consideration of these factors opens up the option of replacing the existing structure. The Populous study for athletic facilities commissioned in 2015 is to include a recommendation for "funding plans." This will be a crucial consideration if Beaver Stadium is to be replaced, because the costs of a new stadium would be enormous. Most new stadiums are built with taxpayer-supported bonds or other forms of public assistance. Given Penn State's past relationships with state government, such an effort seems out of the question; therefore, it is highly unlikely that Penn State will ever be able to afford a completely new stadium of the current capacity.

The future of Beaver Stadium is not just a matter of providing a facility for playing and watching football games. It is an investment that requires careful consideration in today's fiscal climate. But the stadium and its environs also play an essential role in the energy and culture of the University in supporting student spirit and events, cementing alumni loyalty, and providing a stage for institutional development activities.

AFTERWORD: THE POWER OF PLACE

Beaver Stadium still retains its iconic status and conveys a strong sense of place. Football and the stadium are permanent parts of our University culture, but the structure and the experience of players and spectators have both evolved over the decades. The stories of Old and New Beaver Field and Beaver Stadium describe historical change as well as remembered continuity in the presentation and enjoyment of a Penn State home game.

For those who witnessed past games of glory and drama, just seeing the stadium takes them back in time and conjures up vivid memories. But even for those who have only heard the stories of extraordinary Beaver Stadium games, simply being there helps them re-create in their minds the events that took place there.

Of course, the power of the place is not just in our memories; it also prompts our curiosity and enriches our understanding of the history of Penn State football. The overall history of sport in our society is not often accorded the same kind of serious consideration that political, military, or other historical subjects receive. We hope in this book that we have given you, the reader, a sense that there is a history in Beaver Stadium that matters, not just to Penn State football, but to the history of the University and to the role of sport in American higher education.

Penn State football fans love the history, the traditions, and the rituals of attending a game at Beaver Stadium. Penn Staters believe that their school's football ethos is based on ideals that are timeless, that extend into the classroom and go beyond graduation. Those who know the history of Penn State football understand that the foundation of the school's football success was laid down in undefeated seasons of the 1880s through the 1920s, and by outstanding coaches like Bill Hollenback, Hugo Bezdek, and Rip Engle. But in the minds of most Penn State fans today, the character of Penn State football and Beaver Stadium are largely the work of Coach Joe Paterno, his staff, and the hundreds of outstanding players that they recruited and taught. And they appreciate what coaches Bill O'Brien and James Franklin have done to perpetuate that work.

The Joe Paterno era has ended, and the turmoil and pain that surrounded that transition will diminish as time passes. Perhaps some day, as some have proposed, it may be the "Joe Paterno Field at Beaver Stadium," and his statue may stand again. But for Nittany Lion fans, Beaver Stadium will continue to be the "Lair of the Lion" for as long as the memories glow and the future remains bright.

FIGURE 186 Beaver Stadium blue-white stripe-out, Rutgers game, September 19, 2015. Penn State tries out an idea used at a number of other stadiums

NOTES

This book began with a focus on the engineering and construction of the stadium as developed in Professor Harry West's lecture and presentations on the history of the stadium. He relied on the archival files of the Penn State Office of Physical Plant as well as analysis of photographs and design drawings. These were combined with his years of experience as a civil engineer and visits to the building sites and conversations with the engineers and construction managers responsible for the design and construction of the various additions to the stadium. A number of the pictures used to illustrate some of these chapters were taken by him during the various expansion projects.

As the project evolved, the importance of the materials in the Penn State University Archives (hereafter PSUA) for placing the history of Beaver Stadium and its predecessors within the context of the historical development of the university, and the role football has played in that history, expanded. Lee Stout's many years of researching and writing Penn State history, in his work as University Archivist and after, gave him in-depth knowledge of those sources.

Naturally, the three main histories of Penn State were essential to the writing of this book. They are Erwin W. Runkle's *The Pennsylvania State College: 1853–1932, Interpretation and Record* (1933), Wayland F. Dunaway's *History of the Pennsylvania State College* (1946), and Michael Bezilla's *Penn State: An Illustrated History* (1985). Equally important to understanding the engineering and history of the stadium were the PSUA's vast

archives of photographs. Those collections are arranged by subject in photographic vertical files (PVF). The subject divisions of Physical Plant, Athletics, and Students were the most heavily used, along with the portrait vertical files (PorVF) for pictures of individuals.

Other contemporary sources, including the student newspapers the *Free Lance* (1887–1904) and the *Collegian* (variant titles from 1904 to present), were used, as was the student yearbook *La Vie* (1890–present), all of which have been digitized and are available online through the University Libraries' websites. A very useful synopsis of *Collegian* stories arranged by academic year is *The Collegian Chronicles*, edited by Marvin Krasnansky. In addition, the *Students' Handbook of the Pennsylvania State College*, published annually under various titles by the Penn State chapter of the Young Men's Christian Association beginning in 1894, the *Penn State Alumni News*, now the *Penn Stater* (1914–present), football programs (1916–present), media guides (1950–present), and university planning documents (issued occasionally beginning in 1907) all in PSUA, were valuable resources. A variety of archival and manuscript materials including records from the Office of Physical Plant, the Board of Trustees, and the papers of Ridge Riley were also consulted. Personal interviews with administrators and staff involved with the evolution and operations of the stadium were also conducted.

The PSUA provides a portal to a wide variety of materials at https://www.libraries.psu.edu/psul/speccolls/psua.html. These include the

"Penn State History and Traditions" collection, which includes compiled historical data on Penn State presidents, sites dedicated to history and traditions, and access to the digitized version of Bezilla's *Penn State*. Other collections include a wide variety of digitized historical photographs, and the University Park Campus Digital History Archives, which provides many images, maps, and plans for individual buildings. In addition, this is also the easiest way to access the "Historical Daily Collegian Archive, 1887–2010," which includes the *Free Lance*, and the "Digital La Vie (yearbook), 1890–2000."

Details in the various chapters that reference Penn State presidents and other notables come primarily from the published histories by Runkle, Dunaway, and Bezilla, along with articles in their biographical folders (ABVF) in PSUA and on the presidents pages at https://www.libraries.psu.edu/psul/digital/pshistory/presidents.html.

Much of the Penn State football history data, including won-lost records, attendance, coaches' biographies, lists of All-American players, and team rankings of past football seasons, comes from Riley's *Road to Number One*, Prato's *The Penn State Football Encyclopedia*, and Rappaport and Wilner's *Penn State Football*. Additional data have been drawn from Intercollegiate Athletics' websites, particularly the annual football media guides (in PSUA and at https://issuu.com/gopsusports/docs/2015_penn_state_football_yearbook).

Current and historical University data are found on the University Budget Office's webpages, at http://budget.psu.edu/default.aspx, which provides access to budget, tuition, and appropriations figures. The "Penn State Fact Book" provides current and historical data on student enrollments, degrees awarded, and much else at http://budget.psu.edu/factbook/. The current strategic plans and related documents can be found at http://strategicplan.psu.edu/, while the Office of Planning and Assessment has more strategic planning and research studies available at http://www.opia.psu.edu/strategic_planning. The Office of Physical Plant website, http://www.opia.psu.edu/strategic_planning, includes capital plans and building construction information. Similar hardcopy historical information for these offices can be found in PSUA.

Some of the information which might have been specified below on repeated occasions has been summarized above. Where additional sources were used by the authors the materials are identified as much as possible in the order in which they appear in the text.

Finally, both Harry West (now a professor emeritus) and Lee Stout (librarian emeritus) began attending Penn State games as students, West in 1954 and Stout in 1965. They continued to attend games over the following half-century as alumni and faculty members and observed and participated in many of the traditions of football in Beaver Stadium and elements of the game experience. The narrative in chapters 4–12 is thus informed by those experiences.

Chapter 1

The details of the inaugural game in Old Beaver Field come from the December 1887 issue of the *Free Lance*, 70.

For Atherton, see also Williams, *The Origins of Federal Support for Higher Education*.

Beaver has no modern biography, but a number of articles can be found in his biographical folder (ABVF) in PSUA and on the "James Addams Beaver" page of the Pennsylvania Historical and Museum Commission's archived Pennsylvania Governors website, http://www.phmc.state.pa.us/portal/communities/governors/index.html.

A list of trustees and their status on the board through World War II is found in Dunaway, 497–507.

Early student life at Penn State, including sports, is well documented in Runkle, Dunaway, and Bezilla; for a broader overview of student life in general, see Bronner, *Campus Traditions*.

The "Penn State Alumni Association through the Years" timeline, which includes membership milestones, is found on the Association's "About Us/Alumni Association Overview" webpages, at http://alumni.psu.edu/about_us/history.

Espenshade's assessment of athletic favoritism is quoted in Bezilla, 78.

Much more on the evolution of control of athletics can be found in the work of Ron Smith, including both *Sports and Freedom* and *Pay for Play*.

Additional details about the evolution of student interest and involvement in athletics in the 1880s and '90s comes from a variety of articles and notes in the *Free Lance* of those years.

The 1881 "foot ball" game against Bucknell is described in Riley, 11–19. The original set of rules for the game is a section entitled "Revised and Latest Rules of Football; Rugby Union and Association Game," in *Latest Revised Rules for Lacrosse, Football, Ten Pins and Shuffleboard, as Revised and Adapted by the National Lacrosse Association of Canada*, found in PSUA.

The early history of the campus and its buildings is described in Stout, "Penn State's University Park Campus," 1–17, and Zabel, "A Historical Reflection," 43–55.

For more on Captain Charles W. Roberts, see Prato, *100 Things Penn State Fans Should Know*, 182–84. "Camp Roberts" is noted in the *Free Lance*, June 1887, 31.

Budgetary information concerning the construction of Old Beaver Field is drawn from financial reports in the *Annual Report of the Pennsylvania State College for 1887*.

Construction of the "Athletic House" is discussed in the May and June 1892 issues of the *Free Lance*, at 31 and 42 respectively.

Chapter 2

For general historical background of college sports and football in particular, see Smith, *Sports and Freedom*, and Whittingham, *Rites of Autumn*.

Yale's enrollment in 1869 was 736; in 1887 it had 1,245 students while Penn State had grown to only 167. See tables A.1-3 and A.1-4 in Pierson.

For more information on "Pop" Golden, specifically, see Stout's "Penn State Diary" column in *Town & Gown*, December 1999.

The oversight of student organizations and the role of graduate managers are found in two "Penn State Diary" columns in *Town & Gown*, dealing with the history of student newspapers (February 2013) and the Penn State Thespians (November 1997). Particularly valuable for the role of alumni and graduate managers in football was Etter, "From Atherton to Hetzel."

Information about early college football rules can be found in a variety of sources: in this brief overview, Smith, *Sports and Freedom*, Whittingham, and a variety of websites dealing with history of college football rules. An interesting contemporary source is Davis, *Football*.

Stadium capacities were found in Wikipedia, "List of American football stadiums by capacity," https://en.wikipedia.org/wiki/List_ of_American_football_stadiums_by_capacity. Another source that includes discussions of a variety of college stadium histories and traditions is *Sporting News Presents Saturday Shrines*.

Student yells, colors, and related traditions are well covered in numerous university and student publications, particularly student handbooks. Summaries can be found in Runkle, Dunaway, and Bezilla. A short treatment of freshmen scraps and customs is Stout's "Penn State Diary" column in *Town & Gown*, August 1999.

The change of college colors to navy blue and white was voted in by the Athletic Association in April 1890, the *Free Lance*, 306. The story of the pink clothing fading to a dull gray in the sun, or after too many washings, came from George Meek '90, later the graduate manager, but the issue of the unavailability of the proper color of materials is noted in the March 1888 issue of the *Free Lance*, 104.

The evolution of press coverage of Penn State football can be found as appendix C, "The Media Cometh," in Riley, 612–18, and in Prato, *Penn State Football Encyclopedia*, 577–79. The earliest radio "broadcasts" are discussed in Bezilla, and in O'Toole, "Intercollegiate Football and Educational Radio."

The history of Penn State's band and mascot at football games are thoroughly described in Range and Smith, *The Penn State Blue Band*, and Esposito and Herb, *The Nittany Lion*. Information on the "Alma Mater" has been recounted numerous times in a variety of Penn State publications, many of which can be found in the Fred Lewis Pattee ABVF and the Students/Songs and Cheers GVF folders in PSUA. The original cheers cited come from Runkle, 282–83. Unfortunately, there is no compiled history of the cheerleaders, but a

variety of stories appear in the *Collegian, La Vie* (both available online through PSUA website), and other publications.

Information about the Lemont railroad station came from a conversation with Michael Bezilla. Passenger service to campus and to Lemont is discussed in Bezilla and Rudnicki, *Rails to Penn State*.

An account of the 1914 bonfire story appears in the "Penn State Diary" column in *Town & Gown*, August 2013.

Chapter 3

The evolution of college campus architecture and planning is described in Turner, *Campus*. The Penn State master plan of 1907 is detailed in the booklet Lowrie, *A Report and Plan for the Campus*, which can be found in the Administration/Long-Range Planning/Physical Plant GVF folder in PSUA. It is placed in context in Office of Physical Plant Planning and Construction, *History of Master Planning and University Architects*.

The plans for "Pop" Golden's "Great Athletic Park" are summarized in *State Collegian*, January 31, 1907, 1, and November 27, 1909, 9. On Golden in Harrisburg, see Etter, 80–81. The keynote speaker at the 1909 dedication was the Honorable Frank B. McClain, from Lancaster, who as Speaker of the Pennsylvania House of Representatives in 1907–8 played a key role in securing the appropriation. *State Collegian*, May 13, 1909, 1.

The inadequacies of the first press box are quoted from Riley, 612.

Thomas's Emergency Building Fund campaign is also documented in booklets and pamphlets in Administration/Physical Plant vertical files (GVF) and in seven cartons of the records in the Office of Physical Plant, Emergency Building Campaign Records (PSUA 1219, series 7)

The contributions of individual Penn Staters to the war effort are detailed in Alumni Association, *Penn State in the World War*. For more information on Levi Lamb, see the "Penn State Diary" column in *Town & Gown*, July 2011.

Bezilla gives detailed coverage of the de-emphasis of football in the 1930s at 145–49 and 188–93. Riley, 233–79, focuses more on the football team's performance in those years. Hetzel's long statement on the dilemmas engendered by football is quoted from Runkle, 369. A copy of the Carnegie report, Savage's *American College Athletics*, is in PSUA.

The 1930 development plan is summarized for alumni in "A Presentation of Development Plans for Its Building Program 1930," insert to *Penn State Alumni News*, June 1930, also found in the Administration/Long-Range Planning/Physical Plant GVF folder in PSUA. Campus plans for the 1930s and 1940s showing the proposed stadium on the west end of campus come from Office of Physical Plant Planning and Construction, *History of Master Planning and University Architects*.

"Map of the Pennsylvania State College, Past and Present, 1930," by Prof. Andrew Case, was a centerfold insert for *Penn State Alumni News*, January 1930.

Chapter 4

The 1921 planning study was a hand-drawn map entitled "The Pennsylvania State College, A Comprehensive Plan for Campus Arrangement, Proposed by the Dept. of Landscape Design, June 1921, by Arthur Westcott, College Landscape Architect." PSUA Oversize Maps and Charts.

For expansion planning, see "College to Consider Plans for Replacing Beaver Field Stands," *State Collegian*, December 11, 1933, 1. This includes a mention of the expected relocation of the athletic fields to west campus, "where the soccer field and golf course are now located." The sequence of construction is based on design drawings secured from the archives of the University Office of Physical Plant. The dates and size of each phase is clearly enumerated on these drawings, tracing the seating supported by a steel structure from 2,160 in 1934 to 14,700 in 1939. It is on these drawings that the term "Lambert Grandstand" first appeared. Articles that treat the transition and erection of the steel as in New Beaver Field include the Pittsburgh–Des Moines Steel Company brochure, *PDM Steel Deck Grandstands,* in the collections of Prof. West, and Woodbury's *Grandstand and Stadium Design.*

Byron J. Lambert's main patent for "Grand Stand Construction" was US patent 1,452,467, April 17, 1923. This is shown in two images at http://patentimages.storage.googleapis.com /pages/US1452467-0.png (and … -1.png). University of Iowa archivist David McCartney kindly supplied materials on Lambert. An October 30, 1953, article from the *Daily Iowan* marking Lambert's death indicates that this method was used in 1922 to build "10,000 additional seats here on old Iowa field, where university football games once were played. When the present stadium was built, the grandstand was dismantled and set up to serve as balcony seats in the Field House." In a May 11, 1960, article, also from the *Daily Iowan,* an engineering faculty member stated that Lambert was the first to apply structure theory in a folded plate structure. "Folded plating is the pleating of thin material to gain rigidity in various types of structures." Additional Lambert patents (2,180,986 and 2,330,365) were used in the construction of the horseshoe addition and the extension of the east and west stands at New Beaver Field to increase its seating capacity to over 27,000.

For track and field at Penn State, see Hartman, "A History of Intercollegiate Track and Field Athletics." Useful information on dedicated venues and the history of the sport came from Wikipedia, "Track and Field" https://en.wiki-pedia.org/wiki/Track_and_field, and "NCAA Men's Division I Outdoor Track and Field Championships," https://en.wikipedia.org /wiki/NCAA_Men%27s_Division_I_Outdoor _Track_and_Field_Championships, and from "Track and Field History and the Origins of the Sport," http://www.athleticscholarships.net /history-of-track-and-field.htm. For early Penn State Olympians, beginning with Lee J. Talbot who competed in several events in the 1908 London Olympics, see Lucas, *Penn State at the Olympic Games.*

There is no comprehensive history of athletics at Penn State, but *La Vie* yearbooks provide an annual overview of all sports for a given year.

For historical information on the Nittany Lion Inn, see Brannan, Bronstein, and Esposito, *A Tradition of Hospitality and Service.* The Water Tower is not a heavily documented subject, but a *Penn State Collegian* article of September 1, 1938, 3, indicated that the football team would begin to use the first floor of the tower for dressing. Before the early 1950s, the second and third floors had been modified to serve as shower and locker facilities as shown in the "Penn State University Park Campus History Collection," https:// www.libraries.psu.edu/psul/digital/upchc.html, with reproductions of floor plans for the Water Tower at https://collection1.libraries.psu.edu/ cdm/compoundobject/collection/psuimages/ id/4572/rec/1. The Nittany Lion Shrine is thoroughly covered in Esposito and Herb.

A reproduction of the architect's rendering of the projected horseshoe expansion appeared in August 20, 1948, *Centre Daily Times,* copy in the Physical Plant/Beaver Field GVF file in PSUA.

A more detailed story on the expansion is found in an *Alumni News* clipping in the "Beaver Field" section of "Penn State University Park Campus History Collection," https://collection1.libraries .psu.edu/cdm/singleitem/collection/upchc/ id/184/rec/21.

Descriptions of the expansion of New Beaver Field are based on the design drawings from OPP and the decision process in Board of Trustees Building and Grounds Committee minutes for January 24, 1948, and January 22, 1949, and Executive Committee of the Board minutes for December 3, 1948. An article "College Plans Raising Beaver Field Capacity," in the *Daily Collegian,* December 10, 1948, 1, provides an extensive description of planned changes.

Reactions to the de-emphasis program and the efforts of "Casey" Jones are described in Bezilla, and in Riley, 269–76.

Chapter 5

Penn State alum Fred Waring and his Pennsylvanians perform "Collegiate" from a 1925 recording at https://www.youtube.com/ watch?v=fAOLqVcJtD4. In the same year, Waring's Pennsylvanians performed the prologue live performance before the showing of Harold Lloyd's film *The Freshman* at Grauman's Chinese Theatre in Hollywood. The football game sequence, the culmination of the film, can be seen at https:// www.youtube.com/watch?v=yqq15HD1wMs. The game sequences were filmed at the Rose Bowl and the crowd scenes at the 1924 California-Stanford game at Berkeley. The film was a great success and sparked a craze for college films.

Student dress styles are well described in Clemente, *Dress Casual,* which uses Penn State as one of the institutional case studies, and also Bergstein, *Living among Lions,* 35–36. Grace

Holderman's comment on dressing "collegiately well" comes from an interview segment with her in Penn State's sesquicentennial video *Raise the Song*.

Among those who parked cars in College Heights is former athletic director Tim Curley, as he recalled in an interview with the authors, August 4, 2014.

Homecoming weekend arrangements are detailed in *Collegian* and *Alumni News* articles appearing before those games over a variety of years. Houseparty reminiscences as well as Penn game Philadelphia memories come from Ritenour, *"Has Johnny Went Yet?,"* 33, 36.

Arrangements for the Columbia game are detailed in the *Penn State Collegian*, October 23, 1933, 1.

University of Pittsburgh game information comes in part from "Pitt Stadium History" in the 1999 Pitt Football Media Guide, 212–13. University of Pittsburgh Athletics Media Guides collection (1950–2005) at http://digital.library.pitt.edu/cgi-bin/t/text/pageviewer-idx?c=pittathletics&cc=pittathletics&idno=31735038318790&q1=pitt+stadium+history&frm=frameset&view=image&seq=214.

Rail access via the Pennsylvania and Bellefonte Central railroads is discussed in Bezilla and Rudnicki. For bus transportation, see the "Penn State Diary" column in *Town & Gown*, February 2016. Bezilla, 176–78, also describes transportation to and from State College, including Sherm Lutz's pioneering air service and the State College Air Depot. More about Lutz can be gleaned from the Sherman Lutz Collection of papers and photographs at the Centre County Historical Society, 1001 East College Avenue, State College, PA.

The portrayal of State College as a tourist destination in the WPA Guide series was suggested by Thelin, *A History of American Higher Education*, 224.

Chapter 6

In addition to materials in PSUA and the presidents website, the most recent biography of Milton S. Eisenhower is Ambrose and Immerman, *Milton S. Eisenhower*.

Copies of the long-range studies, which are quoted here, can be found in the Administration/Long-Range Planning GVF folder in PSUA. A catalogued copy of Walker's "Penn State's Future—The Job and a Way to Do It" is found in PSUA as well as in the Long-Range Planning folder.

The anticipation of a highway bypass on the eastern side of campus comes from President Eric Walker's report to the Board of Trustees, quoted in Board Executive Committee minutes for October 12, 1956. On December 12, 1958, the Board's Committee on Physical Plant recommended that the executive committee approve granting a right-of-way to the state highway department to use university land to widen East College Avenue between State College and Dale Summit, and to construct University Drive from College Avenue to Park Avenue, passing the western side of the projected stadium.

Discussion of the possible move of the football field to the east campus preceded the "Penn State's Future" planning study. In a February 15, 1956, *Daily Collegian* article, Dean McCoy discussed plans to move many of the practice fields for varsity sports from the southeast corner of the golf course to the east campus. Also in this area would be additional recreational softball fields and tennis courts and a one-story locker-room building for soccer and lacrosse near the ice skating rink.

Dean McCoy's September 24, 1956, letter to S. K. Hostetter is shown as an exhibit in the December 8, 1956, minutes of the Board's Executive Committee. It was at that meeting that the planning study for the relocation was recommended. At the July 25, 1958, meeting the employment of a consulting engineer was authorized to make a complete study and report on the relocation. The comments by "Casey" Jones at the May 23, 1958, Athletic Advisory Board meeting were reported in the June 6, 1958, Board of Trustees minutes.

Michael Baker Inc. submitted its planning study on October 1, and at the October 10, 1958, meeting of the Board's Executive Committee the move of the stadium was authorized, along with the employment of Baker to prepare the detailed plans and specifications, and the direction to the administration to obtain bids for the work. Baker's original engineering report of October 1, 1958, included the details of the two phases of work: to prepare the new site, including a parking lot for 3,500 cars, while carrying out the move after the last football game of 1959.

Board minutes of January 23, 1959, showed acceptance of the various contractors' bids at a total of approximately $1.6 million, with a scheduled completion date of June 15, 1960, which was fulfilled.

The approval of "Beaver Stadium" as the name for the new facility was recorded in the Board minutes of January 30, 1960.

Harry West was on the faculty during the creation of Beaver Stadium, and the accounts of the dismantling of New Beaver Field and the construction of Beaver Stadium in 1959–60 are largely from his recollections.

Chapter 7

Engle's "low-key and curse-free" approach to coaching is quoted from Rappaport and Wilner, 47.

Some of the detailed information on the planning and construction for the lift of the stadium in the 1977–78 time period is derived directly from the photographs used to document this chapter.

In addition, Harry West was on the faculty during this time period, and much of the description of the process is from his recollection, augmented by extensive interactions with W. Herbert Schmidt, associate athletic director during the lift process. West also interviewed Ken Getz, the job superintendent for H. B. Alexander and Sons, Inc., the contractor for the lift about this operation, July 25, 2014.

The comments of Clarence Knudsen and Robert Scannell are quoted from the *Evening News* of Harrisburg, November 17, 1977.

Chapter 8

The story of Penn State's evolution as a nationally recognized university in both academics and athletics is paralleled here by discussion of its economic and political state context. Sources for such changes consulted here were Miller and Pencak, eds., *Pennsylvania*, in particular Philip Jenkins's chapter, "The Postindustrial Age: 1950–2000," 317–70; also the "Overview of Pennsylvania History: 1861–1945, Era of Industrial Ascendency; 1945–2013, Maturity" and the biographical sketches of Pennsylvania governors, both on the website of the Pennsylvania Historical and Museum Commission at http://www.phmc.state.pa.us/portal/communities/.

Joe Paterno's January 22, 1983, speech to the Board of Trustees has been reproduced in several places. The copy consulted here was at http://onwardstate.com/2014/01/22/remembering-joe-paternos-1983-board-of-trustees-speech/.

For more on Penn State and the NCAA Supreme Court decision, see Smith, *Pay for Play*, 139. Also Prato, *Penn State Football Encyclopedia*, 577–79.

The detailed description of the north deck construction and problems comes in part from Harry West's on-scene observations and knowledge of these changes. Robert M. Barnoff, who served as a consultant with Michael Baker Engineers, served as the primary source of information on the deck and the problem with the cracked corbels. Robert DePuy Davis, structural engineer with John C. Haas Associates, was also interviewed regarding the crack problem. W. Herbert Schmidt was a source of additional information.

Franklin Berkey, "Builders Find Cracks in Beaver Stadium," *Daily Collegian*, July 18, 1991, 8; "PSU Officials Decide on Plan for Stadium Repairs," *Daily Collegian*, August 2, 1991, 8; "Repairs Strengthen Damaged Stadium Addition," *Daily Collegian*, August 21, 1991. Phil Kaplan, "Rods to Fix Cracks in the New Deck of Stadium," *Centre Daily Times*, 1991; "Penn State Fans Test New Deck at Stadium," *Centre Daily Times*, 1991; "Ramps Given Safety OK," *Centre Daily Times*, 1991. Clippings of *Centre Daily Times* articles in Harry West collection.

Chapter 9

Information on funding arrangements for buildings and stadium changes are from an interview with Graham Spanier, April 15, 2015.

Penn State's joining the Big Ten Conference is summarized in Prato, *100 Things Penn State Fans Should Know*, 67–70. Also Smith, *Wounded Lions*, 89–99, provides more detail within the larger university context.

Paterno quote on expanding the stadium comes from a news story by Mark Wogenrich, "Standard Bearer for Change," *Allentown Morning Call*, September 25, 2009, in the Joe Paterno ABVF, in PSUA.

Helen Wise quote on the luxury boxes is from "Stadium May Expand to Meet Ticket Demand," *Daily Collegian*, January 23, 1998, 7.

Parking lots have been paved and expanded periodically as the number of structures around Beaver Stadium has increased. See for example, "Stadium West Parking Lot Expansion Is Complete," *Penn State News*, September 3, 2014, http://news.psu.edu/story/324922/2014/09/03/stadium-west-parking-lot-expansion-complete.

During this expansion, the construction photos were taken by Harry West, who had permission to be on the construction site, and he interviewed Mark Battorf and Alan Smith, construction foremen for Barton Malow/Alexander, the contractors for the work. W. Herbert Schmidt was again a prime source of information on the construction, and Bobby White and Brad (Spider) Caldwell provided key information on the operation of the expanded facility. In subsequent years, Bobby White provided tours and opportunities to visit parts of the stadium, and provided Lee Stout with a game-day pass to visit both the luxury boxes and Mt. Nittany Club.

Tom Gibb, "Giant Crane Hoists New Scoreboard into Place Atop Beaver Stadium," Post-Gazette.com, August 9, 2000, http://old.post-gazette.com/regionstate/20000809crane1.asp.

The Big House's growth is documented in University of Michigan, Bentley Historical Library, "The Michigan Stadium Story: Expansion and Renovation, 1928–1997," http://bentley.umich.edu/athdept/stadium/stadtext/stadexp.htm and "Once Again, the Biggest House, 1998," http://bentley.umich.edu/athdept/stadium/stadtext/stad1998.htm.

Chapter 10

The 1996 campus shooting was reviewed in Anna Orso, "Penn State Recognizes 15 Years since HUB Shooting," *Daily Collegian*, September 16, 2011, http://www.collegian.psu.edu/archives/arti cle_433efbb9-0d6c-57ba-bf6f-a6ba5b54cc70.html.

For details of the 2004 Joe Must Go campaign, see Poznanski, *Paterno*, 262–65.

The STEP program is thoroughly covered in the Nittany Lion Club's Seat Transfer and Equity Program website "Pursuing Success with Honor," http://www.gopsusports.com/sports/c-lionclub/ step.html.

The facilities master plan of 2015–16 has been the subject of endless speculation. The study process drew input from many sources; see "Penn State Athletics Selects Populous to Develop Facilities Master Plan," *Penn State News*, October 1, 2015, http://news.psu .edu/story/372916/2015/10/01/athletics/ penn-state-athletics-selects-populous-devel op-facilities-master; Jourdan Rodrigue, "A Look at PSU Facilities Changes: Q&A with Populous Senior Principal Scott Radecic," *Centre Daily Times*, December 16, 2015; "Penn State Athletics Facilities Master Plan Survey Set for Distribution," *Penn State News*, January 11, 2016, http://news.psu.edu/story/386909/2016/01/11/ athletics/penn-state-athletics-facilities-mas ter-plan-survey-set. Kevin Horne includes conceptual illustrations of various improvements in "Potential Beaver Stadium Upgrade Specs Released in Survey," *Onward State*, January 20, 2016, http://onwardstate.com/2016/01/20/ potential-beaver-stadium-upgrade-specs-re leased-in-survey/. Ongoing reports on the plan provide insights or pure speculation, for example, Nate Bauer, "Reduced Capacity Likely for Beaver Stadium's Future, *Blue-White Illustrated*, July 21, 2016, https://bwi.rivals.com/news/reduced-ca pacity-likely-for-beaver-stadium-s-future.

Monument and plaque "General James A. Beaver Monument," http://www.psu.edu/ur/ about/markers/others/jamesbeaver.html. See also *Daily Collegian* articles on its creation, April 28, April 29, and May 8, 1954.

The Penn State historical markers program is described, and the markers listed, at http://www .psu.edu/ur/about/markers/markers.html.

The Penn State All-Sports Museum is described in its information brochure, 2008; and "Penn State All-Sports Museum Marks 10 Years," *Penn State News*, April 19, 2012, http:// news.psu.edu/story/149835/2012/04/19/ penn-state-all-sports-museum-marks-10-years. Lee Stout was also a member of the museum's planning committee and as University Archivist contributed much input to exhibit designs by Hilferty Museum Planning and Exhibit Designs. For further information about the Penn State Sports Archives, from which many objects and illustrations were drawn, see "Penn State Sports Archives Goes Long in Tracking Athletic History," *Penn State News*, October 15, 2013, http://news .psu.edu/story/290975/2013/10/15/athletics/ penn-state-sports-archives-goes-long-tracking -athletic-history.

The Victory Bell class gift of 1979 is described in "Senior Class Gifts: 1861–Present [2012]," at http://giveto.psu.edu/s/1218/images/ editor_documents/Donor_Resources/ Current_Students/Senior%20Class%20Gift/ senior_class_gift_history.pdf. Also see "Bell of the Battleship U.S.S. Pennsylvania," http://www .psu.edu/ur/about/markers/others/bell.html.

Details on the weathervane gift are from Tysen Kendig, "Nittany Lion Weathervane to Prowl Atop Stadium," *Penn State Intercom*, August 9, 2001, at http://www.psu.edu/ur/archives/ intercom_2001/Aug9/vane.html.

The "'Original Nittany Lion' Lands at Penn State All-Sports Museum," *Penn State News*, July 1, 2011, http://news.psu.edu/ story/157005/2011/07/01/original-nittany-li on-lands-penn-state-all-sports-museum. More information about his restoration and installation in Pattee Library can be found in Esposito and Herb.

On the scoreboard logos, see Tim Gilbert, "The History of Penn State's Nittany Lion Logo," *Onward State*, February 7, 2014, http:// onwardstate.com/2014/02/07/the-histo ry-of-penn-states-nittany-lion-logo/#pretty-Photo. Matt Carroll, "Beaming with Pride: Nittany Lion Scoreboard Logo Expected to Soar Thursday at Beaver Stadium," *Centre Daily Times*, May 14, 2014, http://www.centredaily .com/2014/05/14/4178928/nittany-lion-pride -expected-to.html. These were placed on the new scoreboards that replaced those installed in 2000; see Mike Dawson, "High Definition Upgrade for Scoreboard at Beaver Stadium," *Centre Daily Times*, July 12, 2013, http://www.centredaily. com/2013/07/12/3685020/hd-upgrade-for-score board-at-beaver.html.

The installation of the Paterno statue was described in "Statue of Joe Paterno Mounted Outside of Beaver Stadium," November 2, 2001, http://www.gopsusports.com/sports/m-footbl/ spec-rel/110201aaa.html, and documented in photos by Greg Grieco, "Bronze Statue Erected in Honor of Joe Paterno," *Newswire Extra*, 2001 at http://www.psu.edu/ur/extra/2001/paterno/ index.html. For removal of the Paterno statue, see Susan Snyder and Jeff Gammage, "Joe Paterno Statue a Lightning Rod," Philly.com, July 20, 2012, http://www.usatoday.com/story/sports/ ncaaf/bigten/2015/02/05/penn-state-football -joe-paterno-statue/22938085/; Lori Shontz, "Joe Paterno Statue Removed," *Penn Stater*, July 22, 2012, https://pennstatermag.com/2012/07/22/ joe-paterno-statue-removed/.

Joe Paterno's extensive philanthropic contributions to the university are detailed in Smith, *Wounded Lions*, 114–16.

Christopher Phelps quotation is from "Removing Racist Symbols Isn't a Denial of History," *Chronicle of Higher Education*, January 8, 2016, at http://chronicle.com/article/Removing-Racist-Symbols/234862/.

Chapter 11

A good summary of tailgating at Penn State and the role of *Centre Daily Times* editor Jerry Weinstein in its beginning at Beaver Stadium is Prato, *100 Things Penn State Fans Should Know*, 86–90. The reminiscences and stories surrounding tailgating are voluminous, although most books dealing with tailgating are cookbooks. A more serious research-based examination done at Penn State is Deborah Kerstetter et al., "The Multiple Meanings Associated with the Football Tailgating Ritual," *Proceedings of the 2010 Northeastern Recreation Research Symposium, GTR-NRS-P-94*, at http://www.nrs.fs.fed.us/pubs/gtr/gtr-p-94papers/06kerstetter-p94.pdf.

Football parking and traffic control have changed on almost an annual basis and current status can be found on the Penn State Athletics website. A typical article is "Football Parking and Pricing Information for 2015 Announced," *Penn State News*, January 29, 2015, http://news.psu.edu/story/342810/2015/01/29/athletics/football-parking-and-pricing-information-2015-announced.

For Nittanyville, see especially "History of Nittanyville," http://www.nittanyville.com/news_stories/history-of-nittanyville-14826733, and "Rules and Regulations of Nittanyville," http://www.nittanyville.com/news_stories/rules-and-regulations-of-nittanyville-14826658, both on the Nittanyville homepage. See also Mark Viera, "Paternoville Gets a Government," *New York Times*, September 24, 2008, http://

thequad.blogs.nytimes.com/2008/09/24/paternoville-gets-a-government/?_r=0, and Jeff Nelson, "Changes for Paternoville and Student Gate at Beaver Stadium in 2008," *Penn State News*, August 28, 2008, http://news.psu.edu/story/184236/2008/08/28/changes-paternoville-and-student-gate-beaver-stadium-2008.

Information on concessions and the use of volunteers in stands and as hawkers comes from interview with Bob and Eric Byers, July 22, 2014. See also "Video: Go Behind the Scenes with the Beaver Stadium Concessions Team," *Penn State News*, November 11, 2009, http://news.psu.edu/story/172428/2009/11/11/video-go-behind-scenes-beaver-stadium-concessions-team.

Emergency medical systems are detailed in this story, "New Facility Expands Medical Capabilities at Beaver Stadium," *Penn State News*, September 20, 2008, http://news.psu.edu/story/184010/2008/09/20/new-facility-expands-medical-capabilities-beaver-stadium. See also Lauren Ingram, "Penn State Emergency Medical Services Get a Dose of IT: Student EMTs Use Technology to Help Keep Fans Safe at Beaver Stadium," *Penn State IT News*, September 17, 2015, http://news.it.psu.edu/article/penn-state-emergency-medical-services-get-dose-it.

The Blue Band show's evolution is described in Range and Smith. Director Richard Bundy's retirement was the stimulus for a number of reflections on the recent history of the band; see for example John Patishnock, "Bundy Prepares for Last Season Leading Penn State Blue Band," *Centre County Gazette*, September 3, 2014, http://www.statecollege.com/news/local-news/bundy-prepares-for-last-season-leading-penn-state-blue-band,1460625/.

The best synopsis of the evolution of cheering to "We are . . . Penn State," is Prato, "The Origins of We Are!" *BlueWhite Illustrated*, December

26, 2011, https://bwi.rivals.com/content.asp?CID=1310947.

The noise level of the stadium was noted in "Acoustics Team Documents Crowd Noise Effect on Opposition," *Penn State News*, October 29, 2007, http://news.psu.edu/story/192796/2007/10/29/acoustics-team-documents-crowd-noise-effect-opposition. The innovations of Guido D'Elia have received considerable scrutiny. D'Elia's own webpage provided some insights on his role as director of branding and communications, http://gconsultsu.com/guido-delia. His departure from the position engendered several reviews of his impact: Ben Jones, "Penn State Football: Top 5 Creations by Whiteout Creator Sacked by PSU," StateCollege.com, February 27, 2012, http://www.statecollege.com/news/local-news/penn-state-football-top-5-creations-by-whiteout-creator-sacked-by-psu,1013920/, and Bob Flounders, "Goodbye Guido: Penn State Severs Ties with Its Brand Man, Guido D'Elia," PennLive.com, February 29, 2012, http://blog.pennlive.com/bobflounders/2012/02/goodbye_guido_penn_state_sever.html.

The most recent renovations to the press box took place during the summer of 2015; see "Committee Endorses Lasch Building and Beaver Stadium Press Box Renovations," *Penn State News*, March 19, 2015, http://news.psu.edu/story/349063/2015/03/19/athletics/committee-endorses-lasch-building-and-beaver-stadium-press-box.

The alumni *Football Letter* is collected in the PSUA. See also "Nearing Fourth Decade, Black Still Delivers with The Football Letter," *Alumni Insider*, November 26, 2014, http://www.imakenews.com/psaanews/e_article003079380.cfm?x=b11,0,w.

Details of Bob Prince and Mickey Bergstein as radio announcers come from Bergstein, 104, 176.

For Fran Fisher, see "Honorary Alumni Awards: Fran Fisher," http://alumni.psu.edu/awards/individual/honorary/docs/2008-honorary/F.Fisher.pdf ; Connor Whooley, "Former Penn State Football Broadcaster Fran Fisher Dies at 91," *Daily Collegian*, May 14, 2015, http://www.collegian.psu.edu/football/article_af169464-fa60-11e4-9aa3-17ab270bbc79.html, and interview with Steve Jones, June 22, 2015.

Beaver Stadium's turf is a nationally recognized aspect of the stadium. Most of our information comes from interviews with Professor Emeritus A. J. Turgeon, November 16, 2014, and field crew chief Mark Kresovich, September 4, 2014. Professor Turgeon has authored a case study, "Beaver Stadium: A Decision Case in Football Field Management," *Journal of Natural Resources and Life Sciences Education* 28 (1999): 74–78, at https://www.agronomy.org/files/jnrlse/issues/1999/e99-2.pdf. The most recent new turf installation is described at Tony Mancuso, "New Sod Installation Is Underway in Beaver Stadium," *Penn State News*, October 15, 2015, http://news.psu.edu/story/375622/2015/10/15/athletics/new-sod-installation-underway-beaver-stadium. A useful overview of the history of football fields is Jimmy Stamp, "How the Football Field Was Designed, from Hash Marks to Goal Posts," *Smithsonian Magazine*, September 24, 2012, at http://www.smithsonianmag.com/arts-culture/how-the-football-field-was-designed-from-hash-marks-to-goal-posts-48192086/?no-ist.

The 1953 Fordham snow game details are from Bracken, *Nittany Lion Handbook*, 61–62, along with reminiscences of Harry West, who attended the game. Both the 1953 and 1995 snow games are discussed in Prato, *100 Things Penn State Fans Should Know*, 305–9.

Chapter 12

Information on fundraising in the 1950s, primarily the establishment of the Penn State Alumni Fund, Levi Lamb Fund, and the Nittany Lion Club are discussed in Riley, 338, and Bezilla, 249–50.

Commencement details are typically found in the commencement programs housed in PSUA. Lee Stout received his BA in June 1969, in a Beaver Stadium commencement with over 30,000 in attendance on a hot summer day.

Spring Week and other carnival-like activities held around Beaver Stadium are summarized in "Penn State Diary," *Town & Gown*, April 2012.

A brief history of Central PA 4th Fest can be found on its website at http://www.4thfest.org/about-4thfest/vision/.

Civil engineering and surveying camps activities come from the reminiscences of Prof. West and interviews with W. Herb Schmidt.

Information on the use of a stadium elevator shaft for testing came from interviews with Herb Schmidt. See also "University Helping Construct Telescope in Abandoned Beaver Stadium Elevator Shaft," *Daily Collegian*, July 11, 1991, 1.

The wind-tunnel model of the stadium and the results of the study are described in Brad Aris, "Model Shows Punts, Points 'Blowin' in the Wind," *Daily Collegian*, April 23, 1976, 11.

Details of the Trash to Treasure Sale can be found at http://sites.psu.edu/trash2treasure/.

For Quarterback Club background, see Bergstein, 104, 171, and interview with Steve Jones, June 22, 2015.

An NHL game in Beaver Stadium has been a frequent topic of conversation for Penn State hockey fans. The latest expression of interest was cited in Sam Cooper, "NHL Game at Penn State's Beaver Stadium Has Been Discussed," *Yahoo! Sports*, February 26, 2015, http://sports.yahoo.com/blogs/ncaaf-dr-saturday/nhl-game-at-beaver-stadium-has-been-discussed-221413488.html. An actual NHL game will take place at Penn State, but indoors on September 26, 2016: "Penn State, Pegula Ice Arena to Host First NHL Preseason Game," *Penn State News*, June 29, 2016, http://news.psu.edu/story/416170/2016/06/29/athletics/penn-state-pegula-ice-arena-host-first-nhl-preseason-game.

The 2009 Beaver Stadium planning study was shared by Bobby White with Prof. West.

Afterword

The idea of place is a central concern of geography, and in particular cultural geography, where a strong "sense of place" implies a special meaning to a group of people; in this case that strong identity is deeply felt by Penn State students, alumni, and the community of Nittany Lion football fans. The work of geographer Yi-Fu Tuan is a classic entry point to the vast literature on this topic. Numerous historians have also adopted the analysis of sense of place over time to examine the evolving symbolic meanings of battlefields, historic sites, monuments, and other notable places.

SELECTED BIBLIOGRAPHY

Alumni Association, Pennsylvania State College. *Penn State in the World War*. State College: The Alumni Association of the Pennsylvania State College, 1921.

Ambrose, Stephen E., and Richard H. Immerman. *Milton S. Eisenhower, Educational Statesman*. Baltimore: Johns Hopkins University Press, 1983.

Bacon, John U. *Fourth and Long: The Fight for the Soul of College Football*. New York: Simon and Schuster, 2013.

Bergstein, Mickey J. *Living among Lions: Sixty Years with Penn State Personalities, Athletics, and Academics*. Victoria, BC: Trafford Publishing, 2007.

Bezilla, Michael. *Penn State: An Illustrated History*. University Park: Pennsylvania State University Press, 1985.

Bezilla, Michael, and Jack Rudnicki. *Rails to Penn State: The Story of the Bellefonte Central*. Mechanicsburg, PA: Stackpole Books, 2007.

Bracken, Ron. *Nittany Lion Handbook: Stories, Stats and Stuff about Penn State Football*. Wichita, KS: The Wichita Eagle and Beacon Publishing Co., 1996.

Brannan, James, Ben Bronstein, and Jackie Esposito. *A Tradition of Hospitality and Service: The Nittany Lion Inn, 1931–2006*. University Park: Pennsylvania State University, 2006.

Bronner, Simon J. *Campus Traditions: Folklore from the Old-Time College to the Modern Mega-University*. Jackson: University Press of Mississippi, 2012.

Centre Daily Times. *Penn State Forever, 1855–2005: A Photographic History of the First 150 Years*. Battle Ground, WA: Pediment Publishing, 2004.

Chesworth, Jo. *Story of the Century: The Borough of State College, Pennsylvania, 1896–1996*. State College, PA: Borough of State College and the Barash Group, 1996.

Clemente, Deidre. *Dress Casual: How College Students Redefined American Style*. Chapel Hill: University of North Carolina Press, 2014.

Davis, Parke H. *Football: The American Intercollegiate Game*. New York: Scribner's, 1912. Available at https://archive.org/stream/football00232ombp#page/n1/mode/2up.

Dunaway, Wayland Fuller. *History of the Pennsylvania State College*. State College: Pennsylvania State College, 1946.

Esposito, Jackie R., and Steven L. Herb. *The Nittany Lion: An Illustrated Tale*. University Park: Pennsylvania State University Press, 1997.

Etter, Scott C., "From Atherton to Hetzel: A History of Intercollegiate Athletic Control at the Pennsylvania State College, 1887-1930." PhD diss., Pennsylvania State University, 1991.

Hartman, Van A. "A History of Intercollegiate Track and Field Athletics at the Pennsylvania State College." MS thesis, Pennsylvania State College, 1946.

Irwin, Michael, and Joseph Irwin. *Cathedrals of College Football*. Atlanta: Alliance Press, 1999.

Krasnansky, Marvin, ed. *The Collegian Chronicles: A History of Penn State from the Pages of the Daily Collegian, 1887–2006*. State College, PA: The Collegian Alumni Interest Group, 2006.

Kuntz, M. A. *Charlie Atherton: Son of Penn State*. Bloomington, IN: Xlibris Corp., 2005.

Lowrie, Charles N. *A Report and Plan for the Campus and Grounds of the Pennsylvania State College*. State College, PA: Pennsylvania State College, 1907.

Lucas, John A. *Penn State at the Olympic Games (1904 to 1976)*. University Park: College of Health, Physical Education, and Recreation, Pennsylvania State University, 1979.

Mandelbaum, Michael. *The Meaning of Sports: Why Americans Watch Baseball, Football, and Basketball and What They See When They Do*. New York: Public Affairs, 2004.

Miller, Randall M., and William Pencak, eds. *Pennsylvania: A History of the Commonwealth*. University Park and Harrisburg, PA: Pennsylvania State University Press and Pennsylvania Historical and Museum Commission, 2002.

National Lacrosse Association of Canada. *Latest Revised Rules for Lacrosse, Football, Ten Pins and Shuffleboard, as Revised and Adapted by the National Lacrosse Association of Canada*. New York: Peck and Snyder, 1879. In University Archives, MSVF XXX-0056U.

O'Toole, Kathleen M. "Intercollegiate Football and Educational Radio: Three Case Studies in the Commercialization of Sports Broadcasting in the 1920s and 1930s." PhD diss., Pennsylvania State University, 2010.

Office of Physical Plant Planning and Construction, Pennsylvania State University. *History of Master Planning and University Architects.* University Park, PA, 1970.

PDM Steel Deck Grandstands: The Economical Way to Provide Safe Seating for Stadium Audiences. Promotional brochure. Pittsburgh: Pittsburgh–Des Moines Steel Company, n.d.

Penn State Football Yearbook, [current title of annual football guides created for media use]. 2015 issue at https://issuu.com/gopsusports/docs/2015_penn_state_football_yearbook [accessed 7/7/2016. Earlier issues accessed frequently online and in print in the University Archives].

Pierson, George W. *A Yale Book of Numbers: Historical Statistics of the College and University 1701– 1976.* http://oir.yale.edu/1701-1976-yale-book-numbers#A.

Posnanski, Joe. *Paterno.* New York: Simon and Schuster, 2012.

Prato, Lou. *100 Things Penn State Fans Should Know and Do Before They Die.* Chicago: Triumph Books, 2015.

———. *The Penn State Football Encyclopedia.* Champaign, IL: Sports Publishing, 1998.

Raise the Song: The History of Penn State. DVD. Produced and directed by Patrick Mansell. University Park, PA: WPSX-TV, 2005.

Range, Thomas E., II, and Sean Patrick Smith. *The Penn State Blue Band: A Century of Pride and Precision.* University Park: Pennsylvania State University Press, 1999.

Rappaport, Ken, and Barry Wilner. *Penn State Football: The Complete Illustrated History.* Minneapolis, MN: MVP Books, 2009.

Resvine, Dave. *The Opening Kickoff: The Tumultuous Birth of a Football Nation.* Guilford, CT: Lyons Press, 2014.

Riley, Ridge. *Road to Number One: A Personal Chronicle of Penn State Football.* Garden City, NY: Doubleday, 1977.

Ritenour, John P. *"Has Johnny Went Yet?" Reminiscences of John Phillips Ritenour.* Sedona, AZ: Big Park Heights Publishing, 1999.

Runkle, Erwin W. *The Pennsylvania State College: 1853–1932, Interpretation and Record* (1933). State College, PA: Nittany Valley Society, 2013.

Savage, Howard J. *American College Athletics.* Carnegie Foundation for the Advancement of Teaching Bulletin no. 23. New York: Carnegie Foundation for the Advancement of Teaching, 1929.

Smith, Ronald A. *Pay for Play: A History of Big-Time College Athletic Reform.* Urbana: University of Illinois Press, 2011.

———. *Sports and Freedom: The Rise of Big-Time College Athletics.* New York: Oxford University Press, 1988.

———. *Wounded Lions: Joe Paterno, Jerry Sandusky, and the Crises in Penn State Athletics.* Urbana: University of Illinois Press, 2016.

Sporting News Presents Saturday Shrines: College Football's Most Hallowed Grounds. St. Louis: Sporting News, 2005.

Stout, Leon J. "Penn State Diary." Monthly column in *Town & Gown,* 1990–present.

———. "Penn State's University Park Campus: Its Evolution and Growth." In *This Is Penn State: An Insider's Guide to the University Park Campus,* 1–17. University Park: Pennsylvania State University Press, 2006.

Thelin, John R. *A History of American Higher Education.* 2nd ed. Baltimore: Johns Hopkins University Press, 2011.

Turner, Paul Venable. *Campus: An American Planning Tradition.* Cambridge, MA: MIT Press for the Architectural History Foundation, New York, 1984.

Weinreb, Michael. *Season of Saturdays: A History of College Football in 14 Games.* New York: Scribner's, 2014.

Whittingham, Richard. *Rites of Autumn: The Story of College Football.* New York: Free Press, 2001.

Williams, Roger L. *The Origins of Federal Support for Higher Education: George W. Atherton and the Land-Grant College Movement.* University Park: Pennsylvania State University Press, 1991.

Woodbury, William N. *Grandstand and Stadium Design.* New York: American Institute of Steel Construction, 1947.

Zabel, Craig. "A Historical Reflection: The Architecture of Penn State." In *This Is Penn State: An Insider's Guide to the University Park Campus,* 43–55. University Park: Pennsylvania State University Press, 2006.

INDEX

PHOTO CREDITS

Typeset by
Regina Starace

Printed and bound by
Oceanic Graphic International

Composed in
Arno, DIN Next, and Knockout

Printed on
Kinmari Matt

Bound in
JHT